Rainfed Agriculture

Rainfed Agriculture

Authors

R.K. Nanwal

and

G.A. Rajanna

CRC Press is an imprint of the
Taylor & Francis Group, an **informa** business

NEW INDIA PUBLISHING AGENCY
Pitam Pura, New Delhi – 110 088

First published 2021
by CRC Press
2 Park Square, Milton Park, Abingdon, Oxon, OX14 4RN
and by CRC Press
6000 Broken Sound Parkway NW, Suite 300, Boca Raton, FL 33487-2742

© 2021, New India Publishing Agency

CRC Press is an imprint of Informa UK Limited

The right of R.K. Nanwal and G.A. Rajanna to be identified as authors of this work has been asserted by them in accordance with sections 77 and 78 of the Copyright, Designs and Patents Act 1988.

Reasonable efforts have been made to publish reliable data and information, but the author and publisher cannot assume responsibility for the validity of all materials or the consequences of their use. The authors and publishers have attempted to trace the copyright holders of all material reproduced in this publication and apologize to copyright holders if permission to publish in this form has not been obtained. If any copyright material has not been acknowledged please write and let us know so we may rectify in any future reprint.

All rights reserved. No part of this book may be reprinted or reproduced or utilised in any form or by any electronic, mechanical, or other means, now known or hereafter invented, including photocopying and recording, or in any information storage or retrieval system, without permission in writing from the publishers.

For permission to photocopy or use material electronically from this work, access www.copyright.com or contact the Copyright Clearance Center, Inc. (CCC), 222 Rosewood Drive, Danvers, MA 01923, 978-750-8400. For works that are not available on CCC please contact mpkbookspermissions@tandf.co.uk

Trademark notice: Product or corporate names may be trademarks or registered trademarks, and are used only for identification and explanation without intent to infringe.

Print edition not for sale in South Asia (India, Sri Lanka, Nepal, Bangladesh, Pakistan or Bhutan).

British Library Cataloguing-in-Publication Data
A catalogue record for this book is available from the British Library

Library of Congress Cataloging-in-Publication Data
A catalog record has been requested

ISBN: 978-1-032-02460-8 (hbk)

Preface

Rainfed agriculture is practiced in most of the arid and semiarid areas of India. About 67 % of arable land in India (143.2 m ha) is rainfed. In terms of production rainfed agriculture accounts for production of more than 40 per cent of total food grains, nearly 75 per cent of oilseeds, 90 per cent of pulses and 70 per cent of cotton. Most of the rainfed lands are typified by highly fragile natural resource base; the rainfall is low; soils are often coarse textured, sandy, inherently low in fertility, organic matter and water holding capacity; and are easily susceptible to wind and water erosion. Deterioration of natural resources is the main issue threatening sustainable development of rainfed agriculture, more so in the Third World Countries. India will have to produce 300 million tonnes of food grains to feed 1.5 billion populations (approx.) in the coming years. This target cannot be realized from irrigated areas alone as we have irrigation potential for 178 million hectares only. Therefore, we will have to evolve an appropriate technology for dry land farming. On the other hand, we can say that second 'Green Revolution' in Indian agriculture can be in rainfed/dryland agriculture.

In semester system of education the students are quite dynamic for which the students are to be helped for changeover. We can identify their difficulties for comprehensation of language, nonavailability of text books for their semester system. The need for comprehensive information on rainfed farming relevant to undergraduate and postgraduate students of agriculture has been felt for quite sometime. We hope this book will prove a fillip.

The present book suite to the need of students. The chapters of the book have been selected and arranged in such a manner as to lead the students through the entire gamut of rainfed agriculture supported by suitable examples and diagrams. The topics covered are most relevant in view of growing interests in rainfed agriculture technologies. The focus is on new concept and approaches in dryland and rainfed lands; rainfed farming-introduction, characteristics, distribution and problems; efficient management of rainfed crops; water harvesting and moisture conservation; study of mulches and antitranspirants;

principles of intercropping; concept of watershed resource management; drought and its management; soil erosion-definition, nature and extent of erosion; stress physiology; scope of agro-horticultural, agro-forestry and silvi-pasture in dryland agriculture etc.

R.K. Nanwal

G.A. Rajanna

Important Abbreviations Related to Rainfed Agriculture

AICRPDA	:	All India Coordinated Research Project on Dryland Agriculture Hyderabad
AISSLUP	:	All India Soil Survey and Land Use Planning, New Delhi
CAZRI	:	Central Arid Zone Research Institute, Jodhpur
CGWB	:	Central Ground Water Board, New Delhi
CIAE	:	Central Institute of Agricultural Engineering, Bhopal
CIDA	:	Canadian International Development Agency
CRIDA	:	Central Research Institute for Dryland Agriculture, Hyderabad
CSWCRTI	:	Central Soil and Water Conservation Research and Training Institute Dehradun
DDP	:	Desert Development Programme
EGP	:	Effective Growing Period
GIS	:	Geographical Information System
ICARDA	:	International Centre for Agriculture Research on Dry Areas, Aleppo Syria
ICRAF	:	International Centre for Research on Agroforestry, Nairobi, Kenya
ICRISAT	:	International Crops Research Institute for Semi-Arid Tropics Hyderabad
IGFRI	:	Indian Grassland and Fodder Research Institute, Jhansi
IWDP	:	Integrated Watershed Development Project
LCC	:	Land Capability Class
NBSS & LUP	:	National Bureau of Soil Survey and Land Use Planning, Nagpur
NRCAF	:	National Research Centre for Agroforestry, Jhansi
NRSA	:	National Remote Sensing Agency, Hyderabad
NWDB	:	National Wasteland Development Board, New Delhi

viii   Rainfed Agriculture

NWDPRA : National Watershed Development Programme for Rainfed Areas

UNDP : United Nations Development Programme

USDA : United States Department of Agriculture

Contents

Preface v

Important Abbreviations Related to Rainfed Agriculture vii

1. Rainfed Farming: Introduction, Characteristics
   Distribution and Problems 1

2. Efficient Management of Rainfed Crops, Contingent Crop
   Planning for Aberrant Weather Situations 19

3. Management of Rainfed Crops: Choice of Crops and Varieties
   – Planting Methods under Low Rainfall Conditions 47

4. Water Harvesting and Moisture Conservation 67

5. Principles of Intercropping: Cropping Systems/Intercropping
   Systems in Rainfed Agriculture – Mulches and Antitranspirants 85

6. Concept of Watershed Resource Management, Problems
   Approaches and Components 109

7. Drought and Its Management 129

8. Land Use Capability Classification, Scope of Agro-Horticultural
   Agro-Forestry and Silvi-Pasture in Dryland Agriculture 161

9. Tillage, Tilth, Frequency and Depth of Cultivation – Compaction
   in Soil Tillage, Concept of Conservation Tillage 177

10. Soil Erosion: Definition, Nature and Extent of Erosion
    Types of Erosion and Factors Affecting Soil Erosion 197

x   Rainfed Agriculture

11. Drainage Considerations and Agronomic Management Rehabilitation of Abandoned Jhum Lands and Measures to Prevent Soil Erosion 215

12. Stress Physiology: Strategies for Mitigating Stress in Dryland Area 245

13. Tools and Implements Used in Dryland Agriculture 259

14. Collection of Biometric Data on Rainfed Crops and Its Interpretation 269

15. Suggested Readings and Acknowledgement 281

16. Practical Exercises 283

Exercise 1: Rainfall analysis and interpretation 283

Exercise 2: Implements used in dryland agriculture 285

Exercise 3: Agronomic measures of soil and moisture conservation. 289

Exercise 4: Collection of biometric data on crops and its interpretation 290

Exercise 5: Studies on mulches and anti-transpirants 294

Exercise 6: Soil Moisture Determination 397

Exercise 7: Soil moisture determination by neutron moisture meter 300

Exercise 8: Determination of moisture deficit in soil profile 302

Exercise 9: Quantum of water required by plants and on farm soil and water conservation measures 304

Exercise 10: Critical growth stages to water stress for major field crops 306

Exercise 11: Study of cropping systems for dryland areas 307

Exercise 12: Description of dryland areas of Haryana and production/productivity of important dryland crops 309

Exercise 13: Seed treatment, seed germination and crop establishment in relation to soil moisture contents 313

Exercise 14: Study of moisture stress effects and recovery behavior of important crops 317

Exercise 15: Estimation of Moisture index and Aridity index 321

Exercise 16: Collection and interpretation of data for water balance equations 323

Exercise 17: Water use efficiency (WUE) 326

Exercise 18: Preparation of crop plans for different drought conditions ..328

Exercise 19: Visit to dryland research stations and watershed projects. 331

1

Rainfed Farming: Introduction Characteristics, Distribution and Problems

Importance of Rainfed Farming

In India, rain dependent areas are vast and they have a great contribution to make in agricultural production, with just $1/40^{th}$ of the world land. India supports over half of its buffaloes and over one seventh of its cattle and goats that also share land utilization with humans and add to the national wealth. The second largest number (after Africa) of drought victims of the world live in India, out of 143 m ha cultivated land about 43 m ha are under irrigation and the rest of the area (about 70%) is rainfed but all is not too dry. About 90% of it located in the north-west part, out of which 60% is located in Rajasthan and sustains a human population of 20 million and a total livestock population of 23 million. The dry land areas contribute about 42% of food grains, almost all the coarse grains and about 75% of pulses and oil seeds of the total production of the country. About 2/3 of rice and rapeseed mustard and 1/3 of wheat are grown in rainfed areas. A large portion of industrially important crops such as cotton, groundnut and castor are cultivated under dry land conditions and has a great contribution to make to the production of food, fibre, fuel and furniture timber *etc*. Dry land agriculture may be classified into three groups on the basis of annual rainfall.

a) Dry farming : Cultivation of crops in areas where annual rainfall is less then 750 mm and crop failures due to prolonged dry spells during crop period are most common. Dry farming is practiced in arid regions with the help of moisture conservation practices. Alternate land use system is suggested in this region.

b) Dryland farming: Cultivation of crops in areas where annual rainfall is more than 750 mm but less than 1150 mm is called dryland farming. Dry spells may occur, but crop failures are less frequent. Higher evapotranspiration (ET) than the total precipitation is the main reason for moisture deficit in these areas. The soil and moisture conservation measures are the key for dryland farming practice in semi-arid regions. Drainage facility may be required especially in black soils.

c) Rainfed farming: Cultivation of crops in regions where rainfall is more than 1150 mm. There is less chances of crop failures due to dry spells. There is adequate rainfall and drainage becomes the important problem in rainfed farming. This farming is practiced in humid regions.

Steiner et al. (1988) coined aridity index concept of the United Nations conference on desertification based on the balance between precipitations (P) and evapotranspiration (ET) to be appropriate for wide scale adoption. According to this definition, areas with P/ET ratio between 0.03 and 0.20 are arid and areas with 0.2 to 0.5 P/ET ratio are semi arid.

Another influential definition is based on "growing period concept of FAO". Areas having 9 and 74 days are classified as arid and there with a growing period between 75 and 119 days as semi-arid. The growing period is the number of days during a year, when precipitation exceeds half the potential evapotranspiration, plus a period to use an assumed 100 mm of water form excess (or less, if not available) stored in the soil profile.

Indian Council of Agricultural Research (ICAR) while establishing the dryland centers in different agro climatic zones of the country in 1970 used the simple formula of Thornthwaite (1955) for estimating the moisture deficiency index.

$$\text{Moisture deficiency index} = \frac{P - PE}{PE} \times 100$$

Where,

P   = Precipitation, and

PE   = Potential evaporation

According to this classification the limit for semi-arid climates is given by the values of the aridity index varying from -33.3 to 66.7.

Dryland v/s Rainfed farming

Sr. No.	Constituent	Dryland farming	Rainfed farming
1.	Rainfall	< 800 mm	> 800 mm
2.	Moisture availability to the crop	Shortage	Enough
3.	Growing season (days)	< 200	> 200
4.	Growing region	Arid and semi arid areas	Humid and subhumid regions
5.	Cropping system	Single crop or intercropping	Inter cropping or Double cropping
6.	Constraints	Wind and water erosion	Water erosion

Global distribution of Dryland Areas

Arid and semi-arid regions comprise almost 40 per cent of the world land area and are inhabitated by 700 million people. Approximately 60 per cent of these drylands is in developing countries. Low rainfall area constitutes from 75 – 100 per cent of land area in more than 20 countries in the near east Africa and Asia. Farmers in these regions produce more than 50 per cent of the groundnut, 80 per cent of pearlmillet, 70 per cent of gram and 90 per cent of pigeonpea. The yields of these areas are extremely low as compared to those of humid and sub-humid regions. In some countries of sub-Saharan Africa and the neareast, food grain production per capita has declined significantly during the past decade. Although part of this decline can be attributed to high rates of population growth, periodic drought and unfavorable agricultural production and marketing facilities/ policies of the national governments. Much of it results from the steady and continuous degradation of agricultural lands from soil erosion and nutrient depletion and the subsequent loss of soil productivity. Many of these dryland areas are typified by a highly fragile natural resource base. Soils are often coarse textured, sandy and in hereditary low in fertility, organic matter and water holding capacity and easily susceptible to wind and water erosion. Runoff losses during rainfall events commonly exceed 50 per cent. Rainfall pattern are erratic and unpredictable and crops can suffer from moisture deficit and drought even during normal rainfall period. About 75 per cent of the crops produced in the near east region are grown under dryland or rainfed conditions. It is estimated that more than 70 per cent of the projected food and feed deficit by the year 2000 must come from increased yields on established crop lands since very little new arable land is available for agriculture development.

4   Rainfed Agriculture

Table 3 : Extent of non-irrigated lands to food production in selected countries.

Country	Cultivated land Non-irrigated (%)	Total food from non-irrigated lands (%)
China	54	30
India	70	45
Indonesia	66	50
Pakistan	25	18

Global distribution of dryland area

1. Arid and semi-arid regions comprise 40% of the world land area.

2. Some 700 million people inhibit this area.

3. Approx 60% of these drylands are in developing countries.

4. Low rainfed areas constitute from 75-100% of the land area in more than 20 countries in the near east Africa and Asia.

5. These regions produce more than 50% of the groundnut, 80% of the pearlmillet, 90% of chickpea and 95% of the pigeonpea.

6. In India 40% of the total food production comes from drylands which supports 44% of the human population.

Distribution of rainfed areas in India

Three-fourth of our arable land (143 m ha) is rainfed. Food production is tied up with amount and distribution of rainfall; more than 90 per cent of the area under sorghum, groundnut and pulses is rainfed. In case of maize and chickpea, 82 to 85 per cent area is rainfed. Even 78 per cent of cotton area is rainfed. In case of rapeseed/mustard about 65.8 per cent of the area is rainfed. Interestingly, but not surprising 61.7, 44.0 and 35.0 per cent area under rice, barley and wheat, respectively, is rainfed. In terms of production, rainfed agriculture accounts for more than 40 per cent of the total food grains production, nearly 75 per cent of all oilseeds, 90 per cent of the pulses and 70 per cent of cotton.

Area under irrigated and rainfed conditions in India

Class	Area (M/ha)	Total arable land (%)	
Total arable	143.8	—	* includes flood prone area of 15 M/ha
Dryland	34.5	24.0	
Rainfed	65.5*	45.5	
Irrigated	43.8	30.5	

Distribution of low rainfall areas in Haryana and India

In case of Haryana, the districts Sirsa, Hisar, Bhiwani and certain parts of Jind and Mahendergarh districts have an annual rainfall of less than 40 cm. While the districts like Kaithal, remaining parts of Jind, Rohtak, Mahendergarh, Rewari, Sonipat and some parts of Gurgaon, Faridabad, Panipat have annual rainfall of 40-60 cm. The rest of Haryana receives an annual rainfall of > 60 cm. In India the low rainfall areas mostly lies in western and northern India and in central peninsular India.

1. Less than 20 cm rainfall – It includes Jaisalmer and western part of Bikaner districts of Rajasthan. Certain areas of Gilgit and Ladakh districts of Jammu and Kashmir also receives an annual rainfall of < 20 cm.

2. 20-40 cm rainfall – The central parts of Ladakh district of Jammu and Kashmir and Jamnagar & Kutch districts of Gujrat receives annual rainfall of 20-40 cm. The Bathinda and Ferozpur (Punjab), Sirsa, Hisar and Bhiwani (Haryana) and Barmer, Jodhpur, Nagaur, Churu, Sikar, Jhunjhunu, Shri Ganganagar and Hanumangarh (Rajasthan) districts all fall in this category of rainfall.

3. 40-60 cm rainfall-Jalor, Sirohi, Pali, Ajmer, Alwar and Jaipur districts of Rajasthan; Kaithal, Jind, Rohtak, Mahendergarh, Rewari, Sonipat, Gurgaon and Faridabad districts of Haryana; Amritsar, Faridkot, Jallandher, Ludhiana and Kapurthala districts of Punjab; Junagarh, Amreli, Rajkot, Mehsana and Banaskantha districts of Gujarat receives an annual rainfall of 40-60 cm. Similarly South Western Ladakh district of Jammu and Kashmir also receives this rainfall. West Nimar district (Madhya Pradesh); Jalgaon, Aurangabad, Ahmadnagar and Sholapur Districts (Maharashtra); Bijapur, Raichur and Bellary districts (Karnataka); Anantpur and Kurnool districts (Andhra Pradesh) and Periyar and Tirumelveli districts (Tamil Nadu) also receives an annual rainfall of 40-60 cm.

Characteristics of dryland areas

1. Rainfall

The monsoon period is not one of continuous rains in any part of the country. Meghalaya that is uncommon in the behavior of monsoon in the country. There are at least four variations in rainfall.

i)   The commencement of rains may be quite early or considerable delay according to the dates of monsoon for central Maharashtra zone is 10th June but is started as early as 28th May and its onset was delayed upto

6 Rainfed Agriculture

26th June according with the date of onset of monsoon in Rajasthan is between June 21 to 30 but is started as late as up to end of July.

ii) There may be prolonged breaks in rainfall. These breaks in monsoon rains can be different duration. The breaks of 5 to 7 days may not be all serious concern but breaks of longer duration of 2-3 weeks created the situation of plant water stress leading to reduction in agricultural production. Another important point is the physiological properties of soil and available water holding capacity in the relation to the breaks and drought, some of the soils e.g. deep black soils have capacity to store more than 300 mm of water per meter depth where as the desert soil can ensure only 100 mm. The impact of drought is felt more quickly in the soil having less water holding capacity but in soils with high storage capacity 15 days dry break may not affect the crop where as the soils of less storage capacity show wilting symptoms in and dry breaks cannot be predicted for a given season in advance. The drought caused due to this break is therefore of different magnitude and severity.

iii) The rains may terminate considerably earlier or persists long than usual. The problem of withdrawal of monsoon is similar to onset of south-west monsoon. The date of withdrawal of south-west monsoon in Rajasthan is between September 19-20 and its withdrawal was as early as 15th August in some years. The drought of 2000-2001 is due to the early withdrawal of monsoon in Rajasthan.

iv) Uneven distribution of rains: Due to comprehensive analysis relating to the fluctuation of rainfall and soil-water regimes of the different regions of the country. It was found that sub humid climatic regions are prone to both floods and drought. The annual rainfall of India is about 1200 mm. which is slightly more than the global mean of 900 mm but its distribution is uneven. The regional variation is so large that the khari hills of north east get an annual rainfall greater than 6000 mm where as annual rainfall in Rajasthan desert is even less than 150 mm. Significant features of the distribution of rainfall is that a major part of rainfall received from South-West monsoon from June to September. The continuation of this rainfall is quite high or whole of Gujarat and central parts of Rajasthan region the per cent distribution by South west monsoon reduced to 78% in Punjab, about 80-85% in Haryana and North west Rajasthan there is small contribution (6-12%) by winter rainfall.

2. Temperature

Extreme variation in the temperature is the prime characteristic of dryland areas. During the winter season the North India experience very cold temperature which range from -9 to 14°C in the Ladakh region, in cold arid zone to 3-10°C in arid regions of Punjab, Haryana and Rajasthan. Frost often occur in north-west India, Gujarat and Maharashtra experience minimum temperature of the range from 11-15°C, while Karnataka and Andhra Pradesh record 16 to 18°C. The lowest temperature is recorded in January in North India and in December in South India, the temperature beings to rise from March onwards. Desert regions of North-west India record the highest day temperature. May and June are the hottest months and the mean annual temperature is about 48°C. On an individual days temperature of the order of 47 to 50°C are also recorded. Heat waves occurs at maximum temperature in southern part range around 39-40°C and May is the hottest month with the onset of monsoon the mean daily temperature fall sharply from 42°C in June to 33°C in July and 36°C in August to September.

3. Sunshine

Duration of bright sunshine hours is an importance factor in dry areas. Sky remains generally clear favoring in high in salivation by day and rapid back radiation by night. The daily average sunshine hours are 7 to 9 in winter over hour of India, arid zone except cold arid parts where the duration is less than 2.5 hours per day. During the summer season, the arid zones record 9 to 10 hours per day and highest over Rajasthan, Gujarat, Maharashtra regions.

It is less than 8 hours with onset of monsoon, the hours of bright sunshine decreases sharply, 4 to 5 per day in southern arid zone, 4 to 7 hours in northern arid zone in the month of October. When south west monsoon is withdrawal over major part of the country the highest bright sunshine of 10 hours per day are recorded over western Rajasthan with describe gradient south wards and are around 5 to 8 hours in southern arid zone, on an annual basis the northern arid zones recorded slightly higher number of sunshine hours per day (>8 hours) as compared to southern arid zone (7 to 8 hours per day).

4. Wind regime

In these areas wind regime as strong as on a Gujarat followed by Western Rajasthan. Where mean wind speed of >20 km/hour during April to July. Mean wind regime in Punjab, Haryana varying from 2 – 10 km/hour.

8   Rainfed Agriculture

In southern arid zone stronger wind regime (>18 km/hour) prevail in Kurnool region of Andhra Pradesh and over Raichur region of Karnataka during June to August the wind regime is generally low (4 to 10 km/hour) during October to April.

5. Potential evapotranspiration (PET)

Potential evapotranspiration varies from 1 to 2 cm during January, 10 to 14 cm during May to August in cold arid zone. The annual PET in this region varies from 75 to 80 cm in hot arid zone in north western India the annual PET increase from 130 cm in Punjab to 160 cm in Haryana to over 200 cm in Rajasthan. During May-June the PET values at maximum varying from 18-20 cm in Punjab to 22-23 cm in Haryana and 21-31 cm in Rajasthan, during winter the PET values range between 3 to 8 cm. The annual PET values over southern arid zone is 180-185 cm because of the proximate to the equitor and the maximum PE values vary from 22 cm during April-May while lowest during winter season between 10 to 13 cm.

6. Prolonged dry spells during the crop period

7. Low moisture retention capacity

8. Poor soil fertility condition

9. Socio-economic constrains particularly because of the pre-dominance of small and marginal farmers 54% of the holdings is less than one hectare

10. Technological and developmental constraints

11. Limited infrastructure component improper and untimely availability of credits and agricultural inputs

Types of soils in dryland areas, their extent and infiltration characteristics

The resources at the farm level in dryland agriculture are land and rainfall. The drylands have a cafeteria of soil situations. They show great diversity in texture, structure, type of clay, organic matter content and depth. These variations induce significant differences in infiltration rate, erodibility, moisture holding capacity, drainage characteristics, aeration, susceptibility to and recovery from compaction and general response to soil management and manipulation. However, only certain soil types occur widely in the seasonally dry tropics.

1. Vertisols

Vertisols (black soils) are important both agriculturally and ecologically. According to estimates of the National Bureau of Soil Survey and Land Use Planning of India, true vertisols and related black soils occupy 72 million ha in India and constitute 22% of the country's geographical area. Vertisols are predominant in the states of Maharashtra, Madhya Pradesh, Gujarat, Andhra Pradesh and Karnataka. Nearly 84% of Maharashtra, 44% of Gujarat, and 38% of Madhya Pradesh are occupied by these soils.

The annual rainfall ranges from 500 mm. These soils have high clay content (30 to 70%) and high water holding capacity, the water intake capacity early in the monsoon season is high due to deep cracks and larger water retention capacity. The high infiltration rate is further enhanced if the soil management is such that the surface is rough and cloudy and is prepared by the bed and furrow method on a graded contour.

2. Alfisols

Alfisols (red soils) are agriculturally important soils found in drylands. The clay is predominantly kaolinitic in nature. This clay is red or reddish brown to yellowish-brown in color. The plant available moisture storage in the root zone of these soils is usually less than 10 cm. These soils occupy about 20% of the geographical area.

These soils are moderately weathered with a bore saturation of about 80% and dominated by calcium. pH of these soil posses an argillic horizon within the profile. The clay content increases with depth although shallow and gravelly soils also occur due to erosion. They also contain distinct layers of gravel and weathered rock fragments at lower depths, often called 'murram'. The rooting depth of crops is limited by presence of such layers or by compact argillic horizon.

Alfisols in these areas are mostly structure less or massive and hence have low hydraulic conductivity. The structure less nature of these soils in due to the low content of fine particles of clay in the surface horizon, predominance of 1:1 type minerals and cropping systems which add only small amounts of decomposed organic matter.

An important effect of lack of aggregation or presence of unstable aggregation is the tendency of the soils to reduce surface roughness, rapidly seal the surface after rainfall and produce crusting with subsequent drying cycles. The water supply of soils is reduced by impaired infiltration due to lower conductivity of the pores wherever vegetative cover is not available to dissipate the energy of falling raindrops. Although these soils are easy to till when wet, they become hard when dry and are difficult to plough.

Cultivation increases infiltration initially but in the long run porosity and infiltration rates is usually lower than in untilled soil under no mulch cover. Under natural vegetation, erosion is very low. But when these soils are denuded and/or cultivated, high rainfall intensities cause particle detachment and degeneration of infiltration capacity. This results in high runoff and soil loss. Furthermore, alfisols generally posses inherently low water-retention characteristic. Crusted soil surfaces increase the water loss through evaporation.

3. Laterite soils

These soils are well drained with a satisfactory hydraulic conductivity. The clay is mostly either Kaolinitic or illitic type. Laterite soils are generally associated with undulating topography in region with a relatively high annual rainfall. These soils cover 13 M ha in India.

4. Alluvial soils

Alluvial soils are generally loamy sands or sandy loams, very deep with low water holding capacity. Water intake is high and these soils tend to crust on the surface with a beating rain. This leads to runoff during subsequent rains. The moisture and drainage characteristic of the soils are highly variable. These soils occupy nearly 21% of the geographical area.

5. Sierozemic soils

Sierozemic soils are extremely high in sands, loamy sands and sandy loams. Calcium carbonate concretions occur at various depths, influencing the effective soil depth. Surface crusting is a serious problem. Soil erosion through wind is common. These soils are light textured and hence hold less water and nutrients. Subsoil salinity is common due to extreme aridity.

6. Submontane soils

The land is generally undulating and the soils are silty loam to loam. They are medium to deep. Land slide and soil erosion are common. The stored water in these soils may vary from 20 to 30 cm per meter profile. These soils are spread in the hill and foot hill regions of India. Adoption of appropriate soil and water conservation measures is must.

Constraints associated with dryland farming areas

It is a dream desire to eliminate drought from this region. There are many problems that are associated with dryland farming systems. The problems may be grouped into four major factors: climatic, edaphic, technological and socio-economic.

Climatic

1. Scarcity of rain-water which is low and unpredictable with respect to intensity and distribution;

2. High ET (nearly 3650 mm annually compared to total annual precipitation ranging from 350 to 1400 mm);

3. Extreme thermic values (upto 49°C with the mean annual temperature exceeding 18°C at some places);

4. High solar incidence (450 to 500 Cal cm² day⁻¹);

5. High wind velocity with desiccating winds causing a high rate of ET and wind erosion;

6. Low relative humidity;

7. Extensive climatic hazards such as weather aberrations, drought, flood, frost, gale, cyclones and burning winds due to dry, deserted and denuded situations.

Edaphic

1. Poor and marginal lands with soils low in fertility and productivity;

2. Uneven topography with high erodability;

3. Difficulty in workability particularly in vertisols;

4. Shallow or very deep in depth with extreme permeability;

5. Low moisture storage and release capacity particularly in Alfisols;

6. Presence of dissolved injurious salts in ground water;

7. Problem soils with respect to soil reaction (pH) and high concentration of soluble salts in the surface soils;

8. Water logging in level lands; flooding and breaking small field bunds resulting in poor conservation of soil and water;

9. Movement of sand and soil;

10. High surface crusting that leads to poor crop stand and high rate of evaporation and mechanical injury to roots.

Technological

1. Limited choice of crop varieties matching the short moist period suitable for cropping as about 60 per cent of the area has two to four-and-a-half wet months and about 19 per cent of the area has less than two wet months per year;

2. Difficulty in designing a suitable cropping system, crop mixture and crop geometry considering soil type and other flexible components such as the moisture availability index, and the duration of the monsoon;

3. Difficulty in evolving static and adaptable technologies fitting into the prevailing as well as anticipated rainfall conditions on a regional basis. The use of high input based technology is risky;

4. Difficulty in seed production and multiplication specially in an on-far basis;

5. Farming is mainly human and animal power-based which is labour intensive and land saving but costly and time consuming resulting in difficulty in covering the entire arable area under the brief period of favorable weather during the season;

6. Limited use of fertilizers, poor response of biofertilisers and poor availability and the scope to improve it is very restricted;

7. Poor land capacity and the scope to improve it is very restricted;

8. Limited scope of use of residual moisture and nutrients;

9. Difficulty in conserving moisture for proper and timely utilization by crops;

10. Difficulty in reclaiming problems soils;

11. Lacking light and speedy implements with cheap sources of energy;

12. Profuse weed, pest, pathogen and parasite infestation;

13. Unpredicted heavy showers resulting in the accumulation of more than optimum moisture in the soil, impairing field operations including sowing, interculture and harvesting in time;

14. Limited scope of land improvement;

15. Water harvesting and recycling is the high investment proposition at the initial level;

Rainfed Farming: Introduction, Characteristics, Distribution and Problems   13

16. Water damage and loss of run off (up to 50 per cent of incident rainfall) causing loss of soil (greater than 10 t/ha) and formation of gullies resulting in degradation and unsuitability for arable cropping.

Socio-economic

1. Peculiar ecological and socio-economic settings;

2. Frequent failure of crops and unstable production rendering farmers poorer;

3. Low cropping intensity; low farm income, malnutrition and poor quality of drinking water;

4. Small size of farm holdings and high population pressure on land (1.7 persons/ha);

5. Poor quality of produce fetching lower price;

6. Lack of marketing facilities and market incentives;

7. Low level of literacy and poor resource base of the farmers through they are economic minded and rational;

8. Unemployment for most of the period of the year;

9. Farmers are to depend on the favours of the monsoon or on financing institutions which are reluctant to provide assistance as there is a more risk in the recovery of the released amount from farming;

10. Farmers of these regions are deprived of innovations, initiative, inspiration, aspiration and appropriate incentives and appreciations and they wait, watch and worry over their future.

Crop wise production constraints

(A) Bajra

1. Use of lower seed rates and untreated seeds

Farmers must be educated by laying down demonstration on their fields and analyzing the additional costs and additional gains from recommended levels of seed rates as well as treated seeds.

2. Non-adoption of chemical fertilizers including the micronutrients

By laying down demonstration on farmers fields, they must be educated about the use of adequate fertilizers under the rainfed condition.

14   Rainfed Agriculture

3. Thinning and gap filling

The technical feasibility and economic benefits of these practices must be demonstrated to farmers by conducting demonstrations trials on their fields.

4. Non-adoption of sowing by ridger seeder

Tractor drawn ridger seeder is very heavy and costly and there is no equally efficient and good ridger seeder drawn by bullocks.

5. Weed control

To ease the problem of labour shortage, its efficiency should be increased through the use of more efficient implements, tools simple machines; such tools must be developed/popularized on priority basis.

6. Plant protection

Suitable varieties resistant to major insects, pests, and diseases must be evolved. Less costly and adequately effective insecticides/pesticides, weedicides and other plant protection chemicals must be evolved for popularization among the farmers.

7. Lack of varieties for dry farming conditions

There is strong need for developing suitable HYV highly tolerant to moisture stress conditions.

8. Soil moisture

Under the dryland farming situations most of the crops suffer from moisture stress. Appropriate technology for conserving the soil moisture as well as other low cost water harvesting technologies must be evolved and popularized.

(B) Clusterbean (guar)

1. Use of improved varieties

Farmers do not have knowledge about improved varieties that have substantially higher yield potential than the local ones.

2. Use of rhizobium culture

Its utility is shown on paper but most of the farmers are not convinced about it. It needs to be demonstrated on farmer's field to convince them.

3. No/negligible use of fertilizers

The significance of the balance use of fertilizers on yield and profitability of guar needs to be demonstrated by conducting trials on farmer's fields.

4. Inter culture and plant protection operations are not common

The effect of interculture and plant protection operations on yield and profit needs to be demonstrated by the extension agencies to convince the farmers.

(C) Wheat

1. Lower doses of fertilizer application

By conducting demonstrations/trials on farmer's field the economic benefits particularly the additional gains to the farmers due to use of full recommended doses of fertilizers needs to be demonstrated to convince them.

2. Irrigation at right time

Farmers are well convinced about the timely irrigation in wheat, their constraints pertaining to non-availability of adequate electricity/canal water be solved by the Govt. to the maximum possible extent.

3. Effective weed control measures

Farmers are not fully convinced about the magnitudes or increase in yields/ profit by applying adequately weed control measures/interculture operations in wheat.

4. Non-adoption of various plant protection measures recommended by the technical scientists

Most of chemicals recommended for plant protection measures are very costly. Research efforts are needed to find out cheaper and equally effective chemicals for this purpose. Also, most of the chemicals are very costly. Farmers need to be convinced about recommendations concerning the plant protection measures.

5. Non-adoption of weed control measured by majority of the farmers

Farmers need to be educated to convince them about substantial economic benefits of controlling weeds in their fields. Demonstration on farmer's field must be laid out and its economics be worked out to convince them.

(D) Gram

1. Adoption of improved varieties

Relatively superior varieties to common ones are available and well planned yield maximization, demonstration of these recommended varieties needs to be laid out on farmer's field.

2. Use of rhizobium culture

To convince the farmers about the importance of inoculation of seed with rhizobium culture for increasing the yield.

3. Seed treatment for protection against termites

The relevant chemicals must be made easily available to the farmers and they must be educated for treatment of seeds before sowing.

4. Uneven germination and poor plant population due to moisture stress at planting time

The available technology for the conservation of soil moisture must be popularized among the farmers by the extension agencies through laying out field demonstration on the farmer's fields.

5. Very low use of chemical fertilizers

Farmers have apprehension that use of fertilizers under rainfed conditions adversely affects the yield. Also many farmers were not aware of the importance of the phosphatic fertilizers.

(E) Barley

1. Non-use of recommended varieties

In farmer's opinion whatever improved varieties are available are not significantly superior to the locally common varieties to convince the farmers about the superiority of recommended varieties demonstrations be laid out on farmer's field and its comparative economics be worked out.

2. Application of very low dose of chemical fertilizers

Farmers are not fully convinced about the application of true recommended doses of fertilizers under their condition not having adequate irrigation water.

3. Inter culture and weed control measures

Farmers do not adhere to this practice because they are not quite convinced about its economic advantages. They need to be convinced through laying down demonstrations on their fields and working out its comparative economics.

4. Seed treatment against seed borne diseases and termites

Farmers need to be educated about its significant advantages through laying down trials/demonstrations on farmer's field.

(F) Raya (Rapeseed and Mustard)

1. Thinning of excess plant population to make optimum spacing to avoid intra-plant competition

Farmers need to be convinced about its economic advantages by conducting trials on their fields.

2. Application of very low doses of fertilizers

Field demonstrations must be conducted on typical farm conditions to demonstrate the economic benefits of application of the recommended dose of fertilizers as most of the farmers are not quite convinced about these recommendations.

3. Inter culture and weeding

Farmers are not very convinced about its substantial economic gains. It needs to be demonstrated.

4. Insect, pests and disease control measures

Most of the farmers lack technical know-how about various recommended plant protection measures. Demonstrations must be conducted on farmer's field and its comparative economic be worked out to educate them.

2

Efficient Management of Rainfed Crops, Contingent Crop Planning for Aberrant Weather Situations

The conservation of soil and water is very essential for sustainable production, environment preservation and balanced ecosystem. Fifty percent of total geographical area (329 m ha) of our country needs soil and water conservation measures. About 5334 million tonnes of soil is lost annually along with 10 million tonnes of fertilizer and other essential nutrients and organic carbon. Loss of nutrients is almost equal to their total production in India. In India about 70% of the cultivated area falls under the category of dryland conservation measures for soil and water and efficient utilization of these resources is needed in these areas. The fact that rainfed agriculture supports 44% of India's human population and contributes 90 percent of coarse cereals and pulses, 80 percent of oil seeds and 65 percent of cotton and growing realization that further gains in productivity of crops and live stock will emanate from rainfed regions, leave no room for complacency in this regard.

In these areas, rainfall frequently occurs with high intensity that produces higher runoff due to non adoption of soil and water conservation (SWC) measures and poor permeability of the soil. As a result crop grown during rainy season often suffer from moisture stress due to inadequate moisture storage in the soil profile. Secondly soil erosion problems such as sheet, rill and even gully erosion are common in the region.

It has been reported that the black soils of peninsular region of India, under cultivated fallow gives runoff of about 23.4 percent of rainfall and results in a soil loss of 8.94 t ha^{-1} at 1.0 percent slope (Verma *et al.*, 1990). They further reported that the rate of siltation reduced by 83 percent within a span of five years due to adoption of agronomic measure of SWC. These studies clearly

20   Rainfed Agriculture

showed that soil erosion and runoff are the major factors causing land degradation in the region that needs appropriate SWC measures so that crop productivity could be maintained.

Tillage

In dryland, tillage is required, (i) for moisture conservation (ii) to break the hardpan and help root penetration (iii) to control weeds, (iv) provide better aeration and (v) to provide a good seed bed.

Tillage alters the soil physical characters like porosity, bulk density, surface roughness and harness of pans. Different types of tillage implements influence these characteristics in their own way and thus affect runoff and erosion. Tillage operations, such as ploughing, disking, harrowing, planking, cultivating and others are done very carefully so that least loss of soil may take place. Off season tillage is necessary to get a weed free seedbed having good tilth and better moisture conservation. Another, perhaps more important objective is to keep the surface soil loose so that higher rate of infiltration is maintained and runoff reduced. Cultivation of soil with various implements provided as more pore space for storage and increases the amount of water stored. Surface roughness and micro depressions play a great role in high retention of water in soil profile. Profile modification by very deep plough (40-60m depth) with mould board plough in soils with sandy surface but clayey sub surface resulted in better storage at the surface layers. Sub soiling improved soil moisture regimes, reduced runoff and soil erosion and increased yield of barley and soybean by 20 to 30 percent.

Contour cultivation

Farmers generally cultivate their fields along the slope that not only accelerate soil erosion but also adversely affect crops yield. It is therefore, very essential to perform all tillage operations and sowing of crops on contour or across the general slope.

Experiments conducted at Kanpur showed that yield increased with pearlmillet, maize and sorghum with contour cultivation range between 7.5 to 25 per cent. Besides, it reduces runoff by 10 percent and nutrient loss by 40 percent compared to plots with cultivation along slope (Reddy & Reddi, 2002). The results of the experiments conducted at CSWCRTI, Research center, Kota indicate that contour cultivation of sorghum and pigeonpea (1:1) intercropping not only reduced runoff and soil loss but also produced higher yields (Verma *et al.*, 1990).

Graded bunds

Bunds of varying kinds for different purposes and situations have been found effective to check erosion losses from black clay soils. Graded bunds are constructed to control soil erosion on lands having slope in the range of 1 to 6 per cent and the area receiving more than 600 mm of annual rainfall, however, these are also beneficial for deep black soil with rainfall less than 600 mm. These are constructed across the slope on a grade of 0.1 to 0.2 percent spaced at suitable vertical intervals; so as to divide total catchments area into small parts and runoff water from each part is carried away along these bunds without causing any erosion. Graded bunds reduced the runoff from 20 to 48% and soil loss from 2.4 to 4.12 t/ha per year (Singh *et al.*, 1997).

Broad bed and furrow system (BBF)

The BBF system is one of the most efficient land management practices evolved by ICRISAT for *in-situ* moisture conservation in deep black soils. The BBF provide considerable protection against soil erosion throughout the year as they form permanent land feature. Results at ICRISAT indicated that the optimum slope for the bed and furrow system is 0.3 to 0.6% on alfisols (red soils) and 0.4 to 0.8% on vertisols (black cotton soils). The beds function as mini bunds at a grade and help in reducing the velocity of surface runoff and increase in infiltration opportunity time. The excess water is removed through the furrows.

The system involves creation of 90-150 cm wide, 15-20 cm high raised beds with 0.3 to 0.5 percent grades. The beds are stable for 2 to 4 years and conveniently adopted to planting of upland crops in rows spaced at 30, 45, 75 or 150 cm. The system tends to conserve soil, rain water *in-situ* and improves crop yield and sustainability of the production system.

Crop rotation is generally defined as a more or less regularly recurrent succession of different crops on a single piece of land. The crops used are commonly a cultivated crops a small grain and a grass, legumes, or legume grass mixture. Of these, the cultivated crop exposes the soil to the maximum erosion; the small grain allows less erosion and the grass or legume grass crop effectively controls erosion during the period of its life. The principles to be followed in a soil saving rotation are to reduce the time the land is occupied by a cultivated crop as much as the economic situation of the farm will permit, increase as much as possible the time it is covered by legume or a grass, and reduce soil tillage. It is evident that the soil loss from corn following clover in the rotation is less than where corn is grown continuously; also, that the soil loss during the entire rotation occurs mainly while the field is in a cultivated crop.

Soil loss from conventionally tilled continuous cotton crop, average over 70 t/ha annually as compared to previously measured annual soil loss of 19 and 16 t/ha forms soybean and maize, respectively. Pasture has negligible runoff and sediment losses. Inclusion of lucerne in crop rotation reduced soil loss even in soil with 13 percent slope. Soil loss with lucerne was 13.6 t/ha compared with 36.4 t/ha with arable crops over a four year period. Growing a crop that produces the maximum cover reduces runoff and soil loss.

Crops are rotated in order the soil productivity may be preserved and crops yields maintained. Not only does crop rotation serve to reduce soil and water loss, but it also maintains crops yields. Hence, this practice is a profitable one for the farmer to adopt regardless of its value in saving soil and moisture.

Intercropping

Intercropping of short growing legumes in wider inter row spaces of crops provide sufficient cover on the ground there by reducing erosion hazards from the sloping cultivated fields. Prasad and Singh (1994) evaluated the performance of pure castor (inter row spacing 90cm) and castor+green gram (1:2) intercropping on runoff, soil loss and yields of crops on 1 percent slope and found that inclusion of greengram in wider inter row spaces of castor provided adequate cover on the ground and reduced the runoff (17.3 to 21.9) and soil loss (30 to 39%) and increased the total grain production (34.9 to 48.0%) as compared to pure castor. Intercropping on contour not only, reduced the runoff (2 to 3%) and soil loss (5 to 8%) but also resulted in higher yield (48%) (Singh et. al., 1997).

Growing cover crops

Soils bare of vegetation are often prone to erosion. It is also observed that soils low in organic matter content are more easily eroded. Any crop producing dense canopy and foliage to cover the land surface is called as cover crop. Cover crops reduce runoff, add organic matter content to the soil for an improved soil particle binding capacity, protects newly constructed bund/mechanical structures and increase water absorbing capacity and the infiltration of water into the soil. Straight or erect growing crops grown in wide rows with cultivation like cotton, corn, sorghum etc. are highly conducive to soil loss therefore, they must be grown alongwith suitable cover crops like greengram, blackgram, cowpea, groundnut etc. These crops was found satisfactory at 1 per cent slope for reducing soil erosion as these crops covered the land properly within a short span of 30-35 day after sowing (Verma et al., 1990).

Nutrient management

Crop production on eroded soil under dryland conditions is limited not only due to inadequate moisture, but also due to poor fertility status of the soil. Fertilizer application doubled the yield of several crops on eroded soil. Response to fertilizer application is more with soil conservation practices and placement of fertilizers. Drilling of seeds and placement of fertilizer in most subsurface in the same furrow stimulated root growth and extracted more moisture from the deeper layers of soil. Better utilization on soil water as well as nutrients by the crop via better root growth increased grain yield. It was also observed that water holding capacity of alkali soil increased from 42.2 to 53.5 per cent due to addition of organic matter.

Green manuring

Green manuring is the practice of turning under plant material to improve the physical and chemical conditions of the soil. While any thick crop will provide seasonal cover, special crops are grown for green manuring. Sesbania, sunhemp, red clover, sweet clover and alfalfa make excellent green manuring crops when finally turned under, but they are rarely seeded for this particular purpose sunhemp has been widely tried as a covercum green manure crop in *summer/ kharif* season to conserve soil, moisture, improve physical conditions and fertility status of soils a part from controlling a weed growth. The effect of green manuring with sunhemp on runoff, soil loss, fertility improvement, moisture conservation, weed control and crop yield. They found that sunhemp as a green manure crop reduced runoff by 18.6 percent conserved about 8 per cent higher moisture in the upper soil layers, reduced soil loosely 33.1 per cent and enhanced yield of wheat crop by 28 per cent as compared to fallow wheat system.

Mulching

Mulches are used for various reasons, but water conservation and erosion control are undoubtedly the most important for agriculture in dryland (semi arid and arid) regions. While the effectiveness of mulches for water conservation is variable when properly managed. The material include crop residues, leaves, clipping, bark, manure, paper, plastic films, petroleum products, gravel and coal.

Mulching with plant materials reduces soil loss up to 43 times compared to bare soil and 17 times compared to cropped soil without mulches. By increasing the amount of mulch, sediment present in runoff water can be reduced as it covers more soil surface and protects it from rain drop impact. A minimum ground cover with residue should be 30 percent to keep runoff and soil within acceptable limits. Mulches are also applied in narrow slots (vertical mulching) instead of spreading on the soil surface. Crop residues are pressed into harrow continuous slots of 5 to 10 cm width and 20 to 25cm depth. The residue is filled up to 5 cm above the ground level. These slots with residues form a ridge across the slope. They increase infiltration and this reduces runoff.

Use of chemicals

Breakdown of aggregates by the falling raindrops is the main cause of detachment of soil particles. Soils with stable aggregates resist break down and thus resist erosion. Aggregate stability can be increased by spraying chemicals like polyvinyl alcohol at 480 kg/ha, the rate, however, depend on the type of soil. Soil treated with bitumin increase water stable aggregates and infiltration capacity of the soil. However, practical method of increasing stability of aggregates is by application of organic matter, farmyard manures crop residues, green manure *etc.*

Weed control

Weeds (including volunteer crop plants) compete with crops for water, nutrients and light under all cropping conditions. However, under dryland conditions, competition for water generally considered the most important. Consequently, effective weed control before and after crop planting is essential if crops are to produce at their potential under the prevailing environmental conditions. Weed control before planting increases the potential for higher soil water contents at planting, thus providing additional water for the subsequent crop. Weed control during the crops growing season reduces direct competition for water.

Other agronomic practices

Several other agronomic practices helps in soil and water conservation either by increasing infiltration or by providing more opportunity time for infiltration or by improving soil physical properties or by decreasing detachment of soil particles. Application of manure and fertilizer provides early crop cover due to quick growth. Dead furrows (with closed ends) formed at 3.6m interval after emergence of the crop sown across the slope, reduce the length of the run of rain water, hold water and increase opportunity time for infiltration.

Planting geometry refers to the special arrangement of the plants or colonies in the field. These arrangements for a given population can be adjusted either by changing the inter row or intra row distances. The need for suitable geometry under dryland condition is mainly for better utilization of available water. Rectangular planting is found to be more advantageous than square planting for crops raised under conserved moisture situations. When plants are spaced closely with in row and increasing the distance between rows, the soil moisture supply is not exhausted as rapidly as in narrow rows.

It is difficult to predict the season especially under *kharif* rainfall situations. Therefore, it may be advisable to start the season with permissible optimum spacing and optimum population and later on thin out the alternate or third row to provide for wider spacing or by removing the unhealthy seedlings within a row itself if the season is not favourable.

Vegetative control of Guillies

Gullies tend to form wherever runoff is concentrated in unprotected depression channel ways. Once stared, they increase in which rain the land. Where gullies have developed to considerable size, special methods must be used to control their growth and where possible, to reclaim the gullied area for agricultural use. For this purpose, vegetation is most economical and satisfactory, although in some instances, especially in large gullies, the use of close growing plants may need to be supplemented with mechanical measures. Even in small gullies, brush, straw, log, loose rock, or sod dams may be helpful in checking runoff and collecting silt favourable to plant growth while the protective vegetation is becoming established. Gullies often utilized for vegetated water ways. Water must first be diverted, however, the sides smoothed by plowing, and suitable grass or grass legume mixtures seeded. Studies conducted at CSWCRT1, Research centre, Kota reveal that loose boulders and brush wood check dams performed satisfactorily in subsidiary gullies which do not carry much runoff, whereas, loose boulders check dams are suitable for main gullies as they help in silt deposition which facilitate stabilization of gully beds (Verma *et al.*, 1990).

Improved dryland technology for rainfall scarcity areas

Following are the various improved techniques and practices recommended for achieving the objective of increased and stable crop production in dryland areas specially under rainfall scarcity areas.

i. Crop planning: Crop varieties for dryland areas should be of short duration tolerant and high yielding which can be harvested within rainfall periods and have sufficient residual moisture in soil profile for post monsoon cropping.

ii. Planning for weather: Variation in yields and output of the dryland agriculture is due to the aberrations in weather conditions especially rainfall. An aberrant weather can be categorized in three types, *viz.*

26 Rainfed Agriculture

a. Delayed onset of monsoon.

b. Long gaps or breaks in rainfall and

c. Early stoppage of rains towards the end of monsoon season.

Farmers should make some changes in normal cropping schedule for getting some production in place of total crop failure.

iii. Crop substitution: Traditional crops/varieties that are inefficient utilizer of soil moisture, less responsive to production input and potentially low producers should be substituted by more efficient ones.

iv. Cropping systems: Increasing the cropping intensities by using the practice of intercropping and multiple cropping is the way of more efficient utilization of resources. The cropping intensity would depend on the length of growing season, which in turn depends on rainfall pattern and the soil moisture storage capacity of the soil.

v. Fertilizer use: The availability of nutrients is limited in drylands due to the limiting soil moisture. Therefore, application of the fertilizers should be done in furrows below the seed. The use of fertilizers is not only helpful in providing nutrients to crop but also helpful in efficient use of soil moisture. A proper mixture of organic and inorganic fertilisers improves moisture holding capacity of soil and increase tolerance to moisture stress.

vi. Rain water management: Efficient rainwater management can increase agricultural production from dryland areas. Application of compost and farm yard manure and raising legumes add the organic matter to the soil and increase the waterholding capacity. The water, which is not retained by the soil,, flows out as surface runoff. This excess runoff water can be harvested in storing dugout ponds and recycled to donar areas in the severe stress during rainy season or for raising crops during winter.

vii. Watershed management: Watershed management is a approach to optimize the use of land, water and vegetation in a area and thus, to provide solution drought, moderate floods, prevent soil erosion, improve water availability and increase fuel, fodder and agricultural production on a sustained basis.

viii. Alternate land use system: All drylands are not suitable for crop production. Some lands may be suitable for range/ pasture management, for tree farming dryland horticulture, agro forestry systems including alley cropping. All these systems that are alternative to crop production are called as alternate land use systems. This system helps to generate off season employment monocropped dryland and also, minimizes risk, utilizes off season rains, prevents degradation of soils and restores balance in the ecosystem. The different alternate land use systems are alley cropping, agri horticultural systems and silvi pastoral systems, which utilize the resources in better way for increased and stabilized production from drylands.

Efficient management of rainfed crops

Dryland farming refers to high yield and high efficient agricultural production in areas without irrigation capability and mainly depending on natural rainfall through the adoption of a set of dryland farming practices.

Major measures

The objectives for the development of dryland agriculture are as follows: (I) to enhance moisture retention capacity of the soil and retention rainfall by using agronomic, biological and engineering measures in an integrated way; (II) to improve soil fertility and conserve land (soil, water and fertilizer) (III) to change farming practices and make full use of resources like light, heat, soil, fertilizer, water and improved seeds to increase yield and agricultural productivity in the dryland. Major measures include:

Terracing

This measure is meant to build slope land with an angle less than 25° into contour terrace. Slope land with an angle of 10-25° is susceptible to soil erosion due to large angle, steep slope, frequent farming activities and high cultivation coefficient, particularly improper farming practices. Investigation results have shown that farming activities on such slope land have each year led to a soil erosion of 0.43 cm of soil layer and loss of 48 tons of surface fertile soil per ha. To change such slope land into contour terrace, plus other measures like small catchments improvement and biological and agronomic practices, will help to improve production conditions, prevent water and soil erosion and raise land quality and grain yield. Slope land with an angle of 6-10° will be improved to plant crops along the contour using trench culture with 1/2 meter deep and 1 meter wide trench or ridge culture featuring three-dimensional farming approaches with improved soil so as to conserve soil and water, improve soil fertility and facilitate sustainable development.

Furrow drilling

This measure means to use farming practices like moisture retention mulched furrow, machine furrow for drilling and large furrow. The main purpose is to increase active soil layer, improve moisture retention capacity and soil fertility,

Rainfed Agriculture

reduce soil evaporation and improve ecosystem. This practice has shown that can increase soil moisture retention by 81.87 cubic meters per second in places with annual rainfall of 197 mm. This technique is suitable for any slope land with annual rainfall over 400 mm, or for terrace land. When it is used for small angle slope land and terrace land, machines or animals can be employed to reduce labour intensity and speed up engineering progress.

Water cisterns

This measure is meant to build water cisterns to collect rainfall as supplementary irrigation water. In slope land and terrace land, water retention works will be built to collect natural rainfall as supplementary irrigation water for agriculture. In case of serious drought, such water can be used for drip irrigation to increase soil moisture. Due to limited water volume, such technology is usually used together with other water saving measures like wet sowing, plastic mulching, root zone drip irrigation, hole irrigation with mulching so as to enhance crop resistance to drought and ensure stable, high yield. It is applicable to places with annual rainfall less than 350 mm. To build one water cistern with a capacity of 50 cubic meters need an investment of around Rs. 4000. It can provide supplementary irrigation water for 2 ha of land and ensure a yield per ha of over 300 kg. If used in production of cash crops like water melon, greenhouse vegetables and fruit trees, it will bring about even greater yield increase.

Mulching

This refers to mulching with plastic film and crop residue. It can reduce moisture evaporation, increase moisture retention capacity of the soil, alleviate the threat of drought and improve water use efficiency. Plastic film can be used for hole sown wheat, rice, filmside sowing, multiple crops using the same mulching, mulched corn field for water collection with very good results. Wheat farming with plastic film represents a major breakthrough in plastic mulching technology, contributing tremendously to the reform of dryland farming practices and increase of wheat yield. Experiments have shown that wheat with plastic film can produce 30% more yield than conventional wheat drill seeding, increasing yield per ha by 100 kg, net income by over Rs. 1000 and saving water by 100 cubic meters. Mulching with crop residue features easily accessible material low cost of high efficiency, water saving, moisture retention, fertility enhancement, yield increase and no contamination to soil. It can usually improve soil moisture retention rate by 30-50% and increase water supply by 40-80 cubic meters per second. Plus balanced fertilization technology like nitrogen, phosphate and potash replenishment, can at least improve yield per cubic meter by 100 kg, representing an effective solution to the problem of dryness and in fertile soils in dryland region.

Coating

It refers to seed coating mainly using drought resistant agent. The occurrence of drought is often unpredictable. Conventional measures can hardly play their role under such circumstances. Drought resistant chemical agents developed in recent years, like water retention agent, evaporation suppressant and soil regulator, can bring very good results at low cost when used at a time of drought. For instance, seed coated with water retention agent can improve moisture use efficiency by 0.05-0.17 kg/mm; compared with the control group, wheat seed coated with drought resistant agent can increase yield by 20%; and if sprayed before the xerothermic wind, it can increase yield by 9.5-18%.

Fertilization

It refers to the application of chemical fertilizer, organic fertilizer and green fertilizer to improve soil fertility and yield. Besides drought, another factor constraining productivity in dryland farming area is soil infertility. By moisture control through fertilization, we can improve soil resistance to drought through the improvement of soil fertility. The main ways include: 1. To establish rational and effective dryland farming structure, increase the percentage of legume and green fertilizer crops and combine farming with fertility improvement. The key is to adopt a grain crop and summer green fertilizer rotation system or grain oilseed legume rotation system as well as other farming models like grain legume and grain grass intercropping and wheat potato intercropping in order to fully use natural resources like light, heat and water to achieve yield and income increase. 2. To increase fertilizer input by combing organic fertilizer with chemical fertilizer and improve scientific fertilization level. Practices have shown that increased use of chemical fertilizer on dryland can bring about twice as much yield as on irrigated land. Meanwhile, the use of technology like crop residue mulching and quick rotting of crop residue as fertilizer can also increase the application of organic fertilizer and improve the use efficiency of chemical fertilizer.

Deep tillage

This refers to mechanized deep ploughing technology. Through deep ploughing by machines, we can break the sub arable layer without disturbing surface soil layer so as to improve soil ventilation and rainfall retention capacity. This cannot only be used for dryland but also for irrigated land.

Seed replacement

This refers to the use of drought enduring and drought resistant seeds. To use drought enduring varieties is the most cost effective yield increasing technology, at present, most dry areas have their own varieties with good drought resistance. However after such a long time,, most varieties have experienced degradation in their performance and the breeding of new varieties is still lagging behind. It is an urgent need to develop drought resistant varieties and accelerate the purification and rejuvenation process in order to improve yield level.

Crop plan and practices for rainfall scarcity areas

The appropriate crop plan and practices, which have so far been generated and recommended for achieving the objective of increased and stable crop production in dry land areas have been summarized in the following points:-

1. Proper and timely tillage
2. Early sowing
3. Choice of suitable crop/good plant material
4. Proper plant population
5. Cropping system
6. Fertilizer use
7. Timely weed control
8. Efficient rain water storage and its use
9. Alternate land use system
10. Planning for aberrant weather hazard.

1. Proper and timely tillage

Primary tillage is done in order to increase water intake and minimize weed infestation. Farmers do it by country plough. It has to be carried out as early as possible, prior to the sowing season. Keeping this in view, the concept of off-season tillage has been developed. The practice increases soil moisture intake and results in efficient weed control.

Shallow off-season tillage with pre monsoon showers ensures better moisture conservation and lesser weed intensity. In light soils, surface crust formation reduces seedling emergence and rainwater intake, thereby leading to more surface runoff, a light cultivation would be adequate to open such soils. Area with hard pans which tends to develop hard pans due to consistent cultivation and areas with textural profiles i.e. with more clay, as depth increase requires the deep tillage. Deep tillage helps in increasing water in soils having textural profiles and hard pan. For in situ moisture conservation, land has to be opened so that it can cause hurdle to flow of rainwater. Tillage machines of appropriate size and type matching the power sources need to be used.

2. Early sowing

Dry land crops have to be planted at the earliest opportunity after the onset of monsoon. Under some situations, dry sowing is desirable. Early planting helps in better crop yield by providing better moisture conservation, good seedling vigour and availability of longer growing season and avoidance of pests and diseases.

3. Choice of suitable crops/good plant material

Selecting suitable crops and varieties capable of maturing within actual rain fall periods will not only help in enhancing production of a single crop but also in intensifying the cropping intensity. The capacity to produce a fairly good yield under moisture stress condition is the most desirable criteria for selecting a crop variety for dry lands. In other words, crop varieties for dry land areas should be of short duration, drought tolerant and high yielding which can be harvested within rainfall periods and have sufficient residual moisture in soil profile for postmonsoon cropping.The depth of soil, available profile moisture and amount and distribution of rainfall are some of the important factors playing an important role in the choice of crops.

4. Optimum plant population

The optimum plant density or plant population for any given situation results in efficient use of available resources. At this population, production from the entire field if optimized, although any individual plant might produce less than would have occured with unlimited space.

Many factors influence the optimum plant population for a crop are, availablility of water, nutrients and sunlight; length of growing season and potential plant size.

5. Cropping systems

The rainfed areas are generally monocropped having a cropping intensity of 100. Farmers either raise crop in monsoon season or leave the field fallow in post monsoon season or conserve rainwater in monsoon and take a crop in post monsoon season on conserved moisture. However the researchers suggested that the cropping intensity can be increased through inter cropping or double cropping depending upon the weather situation taking the rainfall, soil type and other meteorological parameters into consideration, "effective growing season" have been worked out for different seasons. Based on this information, potential cropping systems have been suggested for the country.

Potential cropping systems in relation to rainfall and soil type

Rainfall (mm)	Soil type	Effective growing season (weeks)	Cropping system
320-600	Red & shallow black soils	20	Single *kharif* cropping
350-600	Deep sierozems and alluviums	20	Single cropping in either *kharif* or *rabi*
350-600	Deep black soils	20	Single rabi cropping
600-750	Red, black and alluvial soil	20-30	Inter cropping
750-900	Alluviums, deep black, deep red and sub mountaneous soils	>30	Double cropping with monitoring
>900	Alluviums, deep black, deep red and sub mountaneous soils	<30	Double cropping assured

Double Cropping

Double cropping is possible in the areas receiving more than 750 mm rainfall with a soil moisture storage capacity of more than 20 cm of available water for a successful double cropping. Early planning of the *kharif* crop and its harvest as early as possible helps in quick establishment of the rabi crop. Most of the *rabi* crops cannot be planted because of thermal sensitivity. The surface soil starts drying up after the harvest of *kharif* crop and seeding of subsequent rabi crop sometimes poses problem. In such situation, pre sowing irrigation for promoting seedling emergence might be important. Use of improved seeding devices like ridger seeder in dry land areas is beneficial.

Double cropping system for Hisar region

Cluster bean—Bajra

Cluster bean—Gram

Inter-cropping

In the seventies, agriculture became more production oriented after the availability of high yielding, short duration and inputs responsive crop cultivars. This developed the concept of inter cropping to produce more over space and time. Keeping this in view, research efforts have been made to identify the most remunerative inter cropping system for different regions in the country. While doing so due importance is given to most common crop of the region and to pulses and oil seeds.

6. Use of fertilizers

Soils of drylands in the country are not only thirsty but hungry also because these soils are severely eroded horizontally as well as vertically whenever efforts are made towards bunding and leveling of fields in dryland areas, it is the surface soil, which is removed. The result and effect is that the fields are rendered shallow in depth and completely deprived of plant nutrients, particularly N, P, K. It is therefore necessary to apply all the three major nutrients in adequate amounts. Fertilizer use plays an important role in drylands. Not only that its use insulates crops against moderate moisture stress as the root system would be more developed compared to unfertilized crops.

Thus fertilizer use helps in more efficient use of moisture, the scarce commodity in dry lands. Since soil moisture is limiting in drylands, the availability of nutrients becomes limited attempt should always be made to apply fertilizer drills drawn by bullock or tractor are available, the objective can be fulfilled. There has been belief among the farmers of dryland areas that use of fertilizer increases the chances of crop failure but recent findings have shown that the use of fertilizer is not only helpful in providing nutrients to crop but also helpful in efficient use of profile soil moisture. In dry land areas, a proper mixing of organic and inorganic material would be desirable. Organic forms have low nutrient content, but help to improve the moisture holding capacity of soils. In addition to yield advantage, nutrients like potassium help to increase drought tolerance by affecting plant soil water relationship. Transpiration losses are reduced and productivity per unit water increases. For the maximum benefit from the applied fertilizer, there is proper placement and split applications of fertilizers need for maximum fertilizer use efficiency, which can achieve.

34   Rainfed Agriculture

Fertilizer recommendation for dry region especially for Hisar region

Crop	Nutrients (kg/ha)		Method of application
	N	P_2O_5	
Bajra	40	20	In 2 splits, ½ drilled at sowing + ½ at knee high stage
Cluster bean	0	30	Placement
Green gram	20	40	All basal drilled
Chick pea	15	40	Drilled
Mustard	40	20	Drilled

Note: 40 P_2O_5 kg/ha to chickpea if grown after bajra. If sown after kharif fallowing adjusts P_2O_5 based on soil test.

7. Timely weed control

Off season tillage, proper seed bed preparation and timely sowing help the crop in combating weeds. Studies conducted in the project very clearly show that the first 25 days period in case of 100 days crops and 35 days period in case of 150 days crops is sensitive to weeds. Therefore, one or two weeding with blade harrow or country plough keeps the weeds well in control. The effects are still better when the practices of inter culture operation with bullock drawn implements along with hand pulling of intra raw weeds is adopted and thus increase crop yields appreciably. For this purpose some implements like blade harrow, rotating weeders like dryland weeder are recommended. Though the herbicides are quite effective in controlling weeds but the high cost of these chemicals takes them beyond the reach of dryland farmers.

8. Rain water management

Two approaches are being adopted in rain water management viz. (a) on farm rain water management and (b) harvesting rain water into dug out farm ponds and recycling on the donor area.

(a) On-farm rain water management

The entry of rainwater into the soil is through the surface. Therefore soil surface need to be more receptive for rainwater. For this we do inter row water harvesting, inter plot water harvesting, sowing across the slopes and use of broad bed & furrows (BBF).

(i) Inter row water harvesting

Cultivation of dry land of crops under ridge furrow system not only helps to take a good crop stand but also helps to conserve soil effectively. In this system, the raindrops absorbed where it falls and does not flow out of the field.

(ii) Inter plot water harvesting

In dryland areas where rainfall (< 300 mm) is hardly adequate to meet water requirement of any of the dryland crops. Crops should be cultivated only in half of the land and the remaining half should be left uncultivated with provision for contributing runoff water to the cropped area. Crops cultivated on this basis are bound to yield fairly even in dry years. This technique is very good for cultivation for arid fruit trees like Ber etc.

(b) Harvesting rain water into dugout farm ponds

Dryland areas do receive atleast one heavy rainfall during monsoon season, which causes runoff particularly from the fields having undulating topography and crusting soil. Runoff in drylands varies from 10-40 percent of rainfall. So effort has to be made to harvest atleast part of this run off in dugout ponds. To check the seepage losses from these ponds several sealing materials have been tried and cheap & effective materials have been identified e.g. in light soils, cement: soil moisture of 1: 8 or plasting with thick coaltar spray on the surface area of the pond are fairly effective.

(c) Recycling of stored water

Even though moisture stress at any crop growth stage can reduce the yield, there are some stages, which are more critical than other. As the water availability is meager in dry lands so it is essential to apply it at right time to get the best return. So the critical stage for irrigation has to be chosen e.g. for rainy season sorghum the flowering stage is most critical and crucial. In tobacco irrigation at 3 weeks before topping had the maximum pay off. For wheat and barley crown root initiation stage is most critical stage. Long duration crops with deeper root systems that responded to higher levels of irrigation compared to short duration shallow rooted crops. This shows that it is important to wet the active root profile for best results.

9. Alternate land use system

All drylands are not suitable for crop production. Some lands may be suitable for range / pasture management, while others for tree farming, ley farming, agro forestry system. All these systems that are alternative to crop production are called as alternate land use system. This system not only helps in generating much needed off season employment in monocropped drylands but also minimize risk, utilizes off season rains which may other wise go waste as run off, prevents degradation of soils and restores balance in ecosystem.

36   Rainfed Agriculture

Crop production may be disastrous in the years of drought where as drought resistant grasses and trees could be remunerative. Scientists of dryland have developed many alternate land use systems which may suit different agro ecological situations. These are alley cropping, agri horticultural system and silvi pastoral systems that utilize the resources in better way for increased and stabilized production from drylands.

10. Planning for aberrant weather conditions

Aberrant weather is a common feature of rainfed agriculture. Four important aberration in the rainfall behaviour have been more commonly observed are:

(i)   The commencement of rains may be quite early or considerably delayed.

(ii)   Dry spell immediately after sowing.

(iii) There may be prolonged breaks during the southwest monsoon season during which most of the dryland crops are grown; and

(iv)   Rains may terminate earlier than normal cessation date or may continue beyond the normal rainy season.

(i) Late onset of monsoon

During some of the years the onset of the southwest monsoon gets delayed so that the crops/varieties which are regularly grown in the region can not be sown in time. Delayed sowing of crop can lead to reduced and even uneconomical crop yields. Under these circumstances, two management options are available.

(a) Transplantation

In case some of the crops, community nursery has to be raised at a point where water is available and transplantation has to be under taken with the receipt of rains. Under delayed conditions, transplanting of pearl millet has been found to be advantageous as the transplanted crop give significantly more grain yield than the direct seeded crop.

(b) Alternate crops/varieties

Certain crops are more efficient either due to the shorter duration or capacity to produce better yields even under relatively unfavourable moisture regimes when sown late in the season. Therefore, crops/varieties have to be chosen depending upon the date of occurrence of sowing rains e.g. in western Rajasthan short location crops like green gram and cow pea were found to perform better than pearl millet under rate sown conditions.

Choice of crop varieties for normal and late sown condition

Crop	Centre	Normal sown	Late sown
Bajra	Hisar	HHB 50, HHB 60, HHB 67	HHB 67
			HHB 94, HHB 117, HHB 197 HHB 216
Mung	Hisar	Muskan, S-9, SML-62, HM-42	Pantmung-I
	Jodhpur	S-8, S-9	
	Varansi	Varsha or K-851	T-44
Cowpea	Hisar	FS-68, HFC-42-1	---
	Jodhpur	FS-68 K-11	
Guar	Hisar	HG-365, HG-75,HG-563	FS-277
Sorghum	Akola	CSH-5, CSH-9	CSH-1
Mustard	Hisar	RH-30, Varuna,RH-781, RH-819	RH-30
Sunflower	Hisar	EC 35737,27250	T-27
Chickpea	Hisar	GNG-146, C-235, HC-1	—

(ii) Dry spell immediately after sowing

During some of the years a dry spell may occur immediately after sowing the crop. This may result in poor germination due to soil crusting, withering of seedlings and poor establishment of crop stand. It is always necessary to maintain proper plant stand to ensure better yields. Therefore, if severe dry spell occurs immediately after sowing it is better to resow the crop than to continue with inadequate plant stand.

(iii) Breaks in monsoon, mid or late season drought

(a) Ratooning and thinning

The water requirement of crop depends upon the solar radiation intercepted by crop canopy. If the drought conditions occur 40-50 days after sowing, the leaf area development will be maximum leading to fast depletion of soil moisture reserve. Therefore reduction in leaf area index by either ratooning or thinning of the crop mitigates the ill effects of drought to a certain extent. Ratooning of a drought affected sorghum crop with a subsequent rain gave 8 q/ha of grain yield over 2 q/ha without ratooning. Ratooning is a high management technique and success depends on the general vigour of the drought affected crop.

(b) Mulching

If the break in monsoon in very brief, soil mulching was found to be a tool in extending the period of storage of water in soil and hence the evaporation losses which in turn leads to extended periods of water availability.

(c) Weeding

Weed control is very essential to save the crop from drought because weeds remove the soil moisture as well as nutrients.

(d) Intercropping and risk distribution

The agricultural drought caused due to prolonged breaks in the monsoon rain can be of different magnitudes and severity and affect different crops in varying degrees. Application of meteorological information in terms of frequency and probability of these breaks can be made to select a combination of crops of different duration in such a way that there is time lag in the occurrence of their growth for appropriate intercropping.

(i) For the areas with uncertain rains in the early part of the season: early planting of deep rooted drought tolerant crops like pigeonpea might be useful followed by planting of other component crop.

(ii) For the areas with uncertain rains in the later part of season: the companion crop should be shorter in duration then the base crop e.g. sorghum, green gram.

(iii) In areas where rainfall is more or less uniformly distributed: it would be ideal to take suitable base crop and a companion crop of either longer or shorterd duration depending on the growing season.

For Hisar region, to minimize the risk the scientists suggested that 60 % area for kharif crops and remaining 40 % should be allocated for rabi crops and for successful kharif system the 50 % area for bajra, 25 % under guar and remaining 25 % area should under short duration pulse/forage crops.

(e) Fertilizer use helps the crop to guard against drought by encouraging development of a root system, which will utilize soil water more efficiently.

(iv) Early withdrawal of monsoon

(i) This type of situation is more dangerous in drought prone areas. Early withdrawal creates difficulties of two types.

(ii) Sowing of rabi crops may be jeopardized (are not sown in early September)

Moisture conserved in the soil is required to be carefully used by appropriate practices.

The suggested practices are as under:

(a) Reduced plant density

Rabi jowar sown during early September at 1.0 to 1.35 lakh plants per hectare. require 50 % reduction. Plant population needs to be adjusted before plants go for their grand period of growth (30-35 DAS).

(b) Use of surface mulch

Moisture can be conserved by using organic surface mulches. For this purpose, mulch material at five tones per hectare is required.

(C) Protective irrigation

If possible protective irrigation may be given. Usually protective irrigation is proposed to be given at 55 to 56 days growth. Due to early withdrawal of monsoon, the same may be applied at 35 to 40 days growth.

(d) Increase frequency of intercultivation

Early stoppage of rainfall brings early cracking in soil. To prevent cracking and there by loss of moisture, frequency of intercultivation may be increase. Usually 3 intercultivation are recommended, if same are increased to 5 or 6, it would be helpftil in creating dust/soil mulch.

(e) Stripping of leaves

It helps to control moisture loss temporarily. This was not found to be useful practice for prolonged drought.

Soil and water conservation measures

Soil conservation is using and managing the land based on the capabilities of the land itself involving application of the best management practices leading to profitable crop production without land degradation.

Water erosion control

Water erosion occurs simultaneously in two steps : detachment of soil particles by falling raindrops and transportation of detached soil particles by flowing water. Therefore, water erosion can be minimized by preventing the detachment of soil particles and their transportation. Principles of water erosion control are:

(i) Maintenance of soil infiltration capacity,

(ii) Soil protection from rainfall,

(iii) Control of surface runoff, and

(iv) Safe disposal of surface runoff.

For a sound soil conservation programme, every piece of land must be used in accordance with land capability classification. Measures of water erosion control can be broadly grouped into three: agronomic, mechanical and forestry measures.

Agronomic measures

In soil and water conservation programme, agronomic measures (conservation agronomy) have to be considered in coordination with others for their effectiveness. These measures are effective on gentle slopes up to 9 per cent. Reduction in runoff is achieved by choice of crops, land preparation, contour cultivation, stripcropping, mulching, application of manures and fertilizers and appropriate cropping systems.

1. Choice of crops: Row crops (erosion permitting crops) such as sorghum, maize, pearlmillet, etc. are not as effective as soil conserving crops (erosion resistant crops) such as cowpea, groundnut, green gram, black gram, etc. Generally, legumes (smothering crops) provide better cover and protection to soil by way of minimising the impact of raindrop and acting as obstruction to runoff.

2. Land preparation: Land preparation including post harvest tillage influence intake rate of water, obstruction to surface flow and consequently the rate of erosion. Deep ploughing or chiselling have been found effective in reducing erosion. Rough cloddy surface is also effective in controlling erosion.

3. Contour cultivation: All the cultural practices such as ploughing, sowing, intercultivation, etc. across the slope reduce the soil and water loss. By ploughing and sowing across the slope, each ridge and plough furrow and each row of the crop act as obstruction to the runoff and provide more time for water to enter into the soil leading to reduced soil and water loss.

4. Strip cropping: It involves growing of few rows of erosion resisting crops and erosion permitting crops in alternate strips on contour with the objective of breaking long strips to prevent soil loss and runoff. Erosion resisting crops reduce the transporting and eroding power of water by obstructing runoff and filtering the sediment from the runoff to retain in the field.

5. Mulching: It reduces soil loss considerably by protecting the soil from direct impact of raindrop and reducing the sediment carried with runoff. A minimum plant residue ground cover of 30 per cent is necessary to keep runoff and soil loss within acceptable limits. As most of the plant remains are used as cattle fodder, availability of organic residues for mulching is the major limitation under conditions of scarce fodder.

6. Organic manures and fertilizers: Organic manures, besides supplying nutrients, improve soil physical conditions. Fertilisers improve vegetative growth (canopy) which aids in erosion control. Improvement in soil structure improves rate of infiltration leading to reduced runoff.

7. Cropping systems: Monocropping of erosion permitting crops accelerate soil and water loss year after year. Intercropping erosion permitting and erosion resistant crops or their rotation have been found effective for soil and water conservation. As. the legumes (cowpea, greengram, horsegram, blackgram) are effective for soil conservation due to their smothering effect, they should be sown in time to develop adequate canopy by the time of peak rate of runoff.

Cover crops such as greengram, blackgram, cowpea, soybean, sunnhemp, groundnut etc restore soil fertility besides reducing soil erosion. Such crops as intercrops with widely spaced crops can give continuous cover of ground protection against erosion.

Forestry measures

Perennial vegetation such as trees and grasses can be successfully used for economic utilisation of degraded lands besides reducing soil and water loss. Effectiveness of perennial vegetation in reducing the impact of raindrop on soil surface and minimising the velocity of runoff of surface flow has been well established.

A large number of tree species were introduced for evaluating their performance under different agroclimatic conditions of the country. Tree species found economical are

Pinus patula (Pinus)

Pinus kesia

Eucalyptus camaldulensis (Eucalyptus)

Moliodora

Leucaena latisilqua (Subabul)

Acacia inearnsii (Acacia)

Acacia nilotica

Acacia tortilis

42 Rainfed Agriculture

The desirable characters of grass for soil and water conservation are perennial nature, drought resistance, rhizomniferous, good canopy, deep root system, prostrate habit and usefulness in cottage industry. Some of the useful grasses and legumes are

Grasses

Cenchrus ciliaris; Chloris guyana; Cynodon dactylon;Dicanthitun annulatuin; Heteropogon contortus; Iseilenia laxtim; Panicunt antidotale; Andropogan haltei; Panicum virgatuni and *Eragrostis curvula*

Legumes

Alylosia scarbaceoides; Centrosenja pubescence and *Siylosanthus hamata*

Hard, medium and software treatments for rainfall scarcity areas

Soil moisture conservation measures coupled with water harvesting help to improve the moisture availability in soil profile and surface water availability for supplemental off season irrigation. The interventions through land management measures have greater role to play on transferring a part of surface water to ground water by recharge. Based on the nature and type of hydraulic barriers and their cost, the conservation measures in arable lands can be divided into following three categories.

1. Hardware treatments

2. Medium software measures

3. Software measures.

1. Hardware measures

These are generally of permanent type, provides for improvement of relief, physiography and drainage feature of the area. These are erected with the major Government support with the purpose to check soil erosion, regulate over land flow and reduced peak flow. At times, they are imposed to completely divert the upland runoff from running into down stream fertile lands. Different hardware measures are explained below.

a) Water ways: Even in drylands, quite often high run off volumes are observed particularly in watershed located in sloppy areas. Such runoff water should, therefore be channalised through a few water ways, as far as possible some of the existing water ways may be developed. Water ways draining larger areas may be designed on hydrologic and hydraulic consideration. No doubt, at place sizable land strips may be found highly advantageous for the survival of many multipurpose useful trees when planted along the water ways. Depending upon land situation alignment of small size water ways may be adjusted with field boundaries; at times, such arrangement are specially useful for safe guarding crops against over stagnation of rain water in black soil areas. Such water course may also be safe generated against erosion by providing mechanical checks and also by raising vegetation which may be useful fodder for animals. In some situation water ways may also have a small section side bund with necessary openings at water entry points.

b) Bunds: These are low height earthen embarkments constructed across the slope in cultivated lands after deciding the location of water ways. The bunds function to intercept runoff, increase infiltration opportunity time and dispose excess rainfall safely.

i) Contour bunds are recommended in dry farming areas with light textured soils of slopes upto 6 per cent and where annual rainfall does not exceed 600 mm.

ii) Graded bunds are constructed in medium to high rainfall areas in poor permeable soil (vertisols) having 2 to 6 per cent slope. These are also quite suitable for the soils having crust formation tendency like red "chalka" soils of Telengana region of Andhra Pradesh.

c) Terracing: cultivated lands having land slope above 10 per cent particularly in Hilly areas should be put under bench terracing by converting the lands into series of platform. The width of bench terrace depends upon the land slope and the permissible depth of the cut. Bench terracing is very much effective in reducing soil erosion in hilly areas.

2. Medium software measures

Medium software are also provided particularly as interbund based treatment, where field sizes are large and conventional bunds are constructed along field boundaries. Such treatments are usually of semi-permanent type and are adopted to minimize the velocity of over land flow. They may need major initiative from cultivators in addition to some grants from the government side. Such measures many lost 2 to 10 years, vegetation component and land configuration may also provide some direct returns on short term basis. However, these need to be modified at time to maintain effectives for erosion control and moisture conservation, and also to minimize risk of providing shelter to harmful pests and around these measures and in certain situations like taking up corrective measures to avoid to much spread of introduced vegetation and being counter productive.

1. **Small section/key line bunds:** A small section bund may be created cross the slope at half of the vertical bund spacing. Such bunds can be nearly 0.1 m² in section; many be renovated at an interval of 2 to 3 years.

2. **Strip leveling:** About 4 to 5 cm strip of land above the bund across the major land slope may be levelled for the purpose. Similarly one or more strips can be created at mid length of slope. Such strips can be erected by running blade harrows after ploughing the field with mould board or disc plough. Such minor range leveling programme may be taken up after every 2 to 4 years.

3. **Live beds:** One or two live beds (2-3 m wide) can be created either on contour or on grade in the inter bund space. The vegetation on the beds may be according to liking of cultivators, it can be annual, perennial or a combination of both.

4. **Vegetative (live barriers):** One or two barriers of close growing grasses or legumes can be created along bunds as well as at the mid length of slope to filter the runoff water or slow down over land flow. Khus could be one of such vegetation. Several other priming grass species that serve as valuable fodder for cattle are being explored as an alternative to khus grass. A miniature bund at lower side of barrier is recommended to help in the development of live barrier particularly in the initial stage.

3. Software treatments

A mention was made that hardware type land treatment all useful for safe runoff disposal and similarly, medium software are essentially to slow down the velocity of over land flow in cultivated fields. However, on several occasion these are found inadequate in attaining equitable moisture distribution for crop growth. In such cases, software treatment are taken up for ensuring uniform soil moisture for satisfactory crop performances. By and large, software treatments are temporary in nature, in that case these are required to be ronade or renovated every year. The entire cost of applying such treatments are to be met by the farmers. Because of favourable economics, a few of these treatments have gained wide acceptances in the recent years.

1. **Contour farming:** Contour farming is one of the easiest and most effective and low cost method of controlling erosion and conserving moisture. With contour farming, tillage operation are required / carried out along contours. It creates numerous ridge and furrows for harvesting sizable amount of runoff inside the soil.

2. **Compartmental bunds:** Compartmental bunds, converting the area into square / rectangular parcels – are useful for temporary impounding of water for improving moisture status of the soil. These are made using bund former. In a medium, deep black soils, they are found advantageous in storing the rainfall received during the season in soil profile thereby augmenting the soil moisture for use by rabi crops. The size of the compartments may be fixed considering the size of the interbunded land.

3. **Scoop formation:** Making small depression on soil surface. This practice can reduce the runoff of water by 69 per cent and soil loss by 53%.

4. **Tile ridges:** Ridges and furrow system helps in moisture increment in soil profile and can reduce the runoff. In tile ridges we can reduce runoff by 39 per cent and soil loss by 28%.

5. **Dead furrows:** Dead ferrows are laid across the land slope in rolling lands to intercept the runoff. The spacing between dead ferrows between 2 to 5 metre or 4 to 7 crop rows. This system works well in alfisols.

6. **Broad bed and furrows:** Broad bed and ferrous system implies shaping alternative bed and ferrows. This technique is especially suited to black soils. Where crops are sown on preformed bed. This systems is made before the season and is maintained year after year. The planting is done on the bed. Generally the depth of each furrow is kept 0.15 metre and the inter furrow spacing is maintained at 1.5 metre.

7. **Tillage:** Tillage operations help in rain water infiltration. Off season tillage, in particularly has been found quite useful in most rainfed areas. Tillage operations make the soil receptive to rainfall. This practice is very useful in light soils often prone to crusting.

8. **Mulching:** Surface mulching protects soil against beating action of rain drop and also it increased water infiltration into the soil. Further it help in minimizing water evaporation from the soil surface. Some times dry soil mulch created simply by stirring the soil has been found effective for good performance.

9. **Deep ploughing during summer:** Deep ploughing during summer particularly in soils with slowly permeable sub soil layer has been found to increase moisture storage in the root zone and consequently the yield. Scientists reported that deep ploughing to 30 cm where sub soil had slowly permeable layer at Ambala and Ludhiana increased the moisture storage in soil and enhanced wheat yield by 14%.

46   Rainfed Agriculture

10. **Inter row water harvesting:** In low rainfall areas the water is collected in already opened furrows rather than every distributing the rainfall in order to increase moisture content for raising crops. This technique is also useful or used to drain excess water and to provide aeration to growing crops in high rainfall areas. Water stored in furrow is easily available to crops. Experiments at International Crop Research Institute for Semi arid Tropics and under AICRP on Dryland have indicated that inter row water harvesting is quite useful for raising yield level in semi-arid area.

11. **Alternative planting pattern:** In areas where plants often experience moisture stress, particularly during long dry spell, the techniques of planting 'narrow' ridge furrow method has been found beneficial in making good use of water or rain water. In this system kharif crops are planted on ridges and rabi crops planted in ferrows subsequently in the winter season. Ridges serve as micro-catchment for furrow where rainfall accumulates. Water collected in furrows meet the water need at crucial period of plant life and also ensures better aeration for crop planted on ridges during the kharif seasons. After harvesting of kharif crops, rabi crops are planted in furrow where sufficient moisture have been stored for germination of rabi crops.

Inter plot water harvesting: In drought pone areas, it may not be only feasible but desirable to leave a part of land for water harvesting and a part for cropping. The first portion, which is relatively elevated, is used for conveying water is runoff plot and the other one which receive the runoff water (relatively low land) is called runoff plot. Experiment at Jodhpur has shown that the total production by cropping only $2/3^{rd}$ of the field (leaving one third) by adopting runoff farming is the same as obtained from cropping the whole field.

3

Management of Rainfed Crops Choice of Crops and Varieties Planting Methods Under Low Rainfall Conditions

Soil and moisture conservation measures are a prerequisite for ensuring a good crop stand, growth and development of crop, and higher crop yields. These measures depend mainly on the land (topography and slope), soil type, rainfall pattern, and nature of the crop. The various management practices are:

Tillage

Off-season tillage is necessary to get a weed free seedbed having good tilth and better moisture conservation. Pre-monsoon showers should be utilized for deep ploughing (where necessary) and blade harrowing. This will enable sowing of large areas in limited time- a necessary in dryland areas. Higher yield advantage with deep ploughing (22 cm) has been obtained in case of sorghum, pearl millet, maize, and rice. In case of rice, use of mould board plough to a depth of 30 cm every year increased the rice yield at Ranchi. For small seed crops like finger millet, soil compaction below the seeding zone is absolutely necessary for getting good crop stand establishment and higher yields. Tillage requirement for pulse crops (cowpea, mungbean, black gram) varies with the type of soil on which they are grown. For loamy sand or sandy loam soils, one or two cross cultivation using a sweep harrow, will be enough to provide a good tilth.

In medium to heavy textured soils, summer cultivation with a disc plough followed by a cross-disc harrow will be adequate for obtaining a good seedbed. The field preparation for pigeonpea normally consists of one cultivation with soil turning

48   Rainfed Agriculture

plough during summer, utilizing premonsoon showers, followed by a cross harrowing or ploughing with a country plough on or before onset of monsoon, and finally planking. Groundnut requires a loose and friable seedbed for the pegs to penetrate easily and also prevent loss of nuts during harvesting. For sesamum and rapeseed mustard, a well prepared seedbed is required for uniform plant stand.

An ideal seedbed for cotton is formed if the soil is loosened and pulverized to a depth of 10 to 12 cm, kept free of weeds and compacted by an inverted blade harrow before sowing. Deep ploughing is not necessary except on soils carrying weeds like *Cynandon dactylon* and *Cyperus rotundus*. In black soils, ploughing once in three or four years, keeps the weed intensity down. Good initial surface drainage and prevention of water logging is essential.

Wheat in dryland areas, usually grown after fallow, needs timely tillage operations so as to get the best advantage, of limited rainwater. More importantly, tillage operations are to be carried out at optimum soil moisture conditions. One mould board ploughing plus two harrowings can achieve a clod size of 6 to 8 mm in black soil region of Madhya Pradesh, provided ploughing is done at 23 per cent surface soil moisture conditions. Barley does not need as fine a seedbed as wheat. One ploughing with soil turning plough following by leveling and rolling would be good enough.

For *rabi* sorghum, usually grown on black soils, there is no need for deep tillage. Harrow will accomplish a good seedbed. However, intercultural operations are necessary to keep the weeds under check, close up cracks, and conserve moisture. Compaction after seeding was found advantageous, resulting in improved emergence of seedling (by more than 50% of the control), early seedling vigour, and increased root growth. The advantage of compaction was more pronounced in years of low rainfall after seedling than in the normal years. Rapeseed and mustard is grown in the post rainy season on conserved soil moisture and, therefore, favourable soil moisture conditions in the seeding zone at sowing time are necessary. Towards the end of the rainy season each ploughing operation must be followed by planking so that soil becomes pulverized and favourable moisture conditions are created.

Seeding

Sub-optimal and gappy plant stand is largely responsible for low productivity of dryland crops. Proper germination and good crop stand is the first requirement for obtaining higher yields. It is also necessary for making an efficient use of inputs-seeds, fertilizers, *etc.* on one hand, and labour, time, and money, on the other. Besides soil tilth which has to be good and firm with adequate moisture in

Management of Rainfed Crops 49

the seedling zone, seedling methods and time of sowing exert significant influence on crop stand establishment. Seed viability and seed health are yet other factors to reckon with.

Seeding methods

Upland direct-sown rice is sown either in dry (friable) seedbed or in puddle soil conditions. The seed is broadcast or line sown. Pregerminated or sprouted seeds are usually used in case of puddled conditions. Superiority of seeding upland rice with seed drill has been clearly established on dry seedbed. The use of seed cum fertilizer drill enables the placement of the seeds in moist zone and at an appropriate depth, ensuring good crop stand establishment. A country plough traditionally sows almost all dryland crops. As an alternative to a seed drill, a pora (bamboo/aluminium/tin) funnel is tied to the handle of the country plough, and seeds are drilled through a bowl provided at the head of the funnel. This enables placement of the seeds in the moist zone. However, because of hand metering of seeds, crop stand is not as uniform as in case of crop sown with a seed drill. In case of crops sown during the monsoon season, kera method of sowing is good enough. In this method, seeds are drilled behind the plough in furrows. Seeded furrows are usually covered by planking. In the para method of sowing, seeded furrows are not covered.

Seeding depth

Depth of seeding has been shown to be critical for the establishment of satisfactory stand of pearl millet, finger millet and other millets, sesamum, and rapeseed mustard. Seeding depth of 2 to 3 cm is about the optimum for these crops. For maize, sorghum, wheat, barley, pulse crops and most of the oilseed crops, 5 to 6 cm has been found to be the optimum seeding depth. In light textured soils, castor can be sown a little deeper. Soybean should been sown at 3 to 5 cm depth. Finger millet is often broadcast with heavy seed rate (25 kg/ha) and thinned out later by cross cultivation with fine hoes. This results in non-uniform stand and too much of population. Sesamum seed being small, it should be mixed with dry earth, sand or well sieved farm yard manure to increase the bulk and to ensure even distribution of seeds while sowing. Drill sowing is by far the best, whatever the crop.

Transplanting

Transplanting has been found to be advantageous, especially under delayed sown conditions. The transplanted crop of pearl millet gave significantly more grain yield than the direct seeded crop. In case of finger millet, 20 to 25 days old seedlings of a medium duration variety are ideal for transplanting. As a thumb rule, the seedling could be let in the nursery for a week for every month of its total life period.

Seeding time

All rainy-season crops, including upland rice, should be sown after the first soaking rains when the top 22 to 30 cm soil layer is thoroughly wetted to sustain germination, emergence and early seedling growth for atleast 15 to 20 days. Such conditions normally arrive by the end of June or beginning of July, in most parts of the country. In case of sorghum, progressive delay of 7, 14 and 21 days cause substantial reduction in crop productivity, the average reduction in grain yield being of the order of 319 to 370 kg/ha/day. The reduction in grain yield was attributed to decrease in number of seeds per panicle and seed weight. Late sown crop attached more pest attack.

Time of sowing is crucial for obtaining higher yield of all dryland crops. Sowing of pearl millet at first monsoon showers was found to be the best. As a dryland crop, finger millet is traditionally sown in July in the Southern states and in May-June in the Northern states. Some of the short duration varieties can be sown late but their yields are low. Sowing long duration varieties late in the season reduced their duration as well as yield. Optimum sowing time for foxtail millet is August-September in Tamil Nadu, July in Karnataka and first fortnight of July in Andhra Pradesh. For plains of Uttar Pradesh, late may to middle of June is the optimum sowing time. Kodo and Proso millet are sown with the onset of monsoon, preferably in July.

System of planting

System of planting is known to exert considerable influence on yield of certain crops, particularly in normal and subnormal years. In case of post rainy season sorghum, paired row planting showed distinct advantage in grain production compared to normally recommended spacing of 45 cm.

Row spacing and plant density

The inter- and intra-row spacing for *kharif* crops in general is 45 x 10 cm and for *rabi* crops in 30 x 10 cm. For optimum plant density, precision in spacing and seed rate is must.

Weed management

In dryland areas, weeds compete with crops for nutrients in the early stages of crop growth, and later for moisture and light. The competition for moisture becomes critical with increasing soil moisture stress. Further, for producing equal amount of dry matter weeds transpires more water than do most of crop plants.

In general, 4-6 weeks period is more critical from competition point of view in almost all the major crops. It is important to know the critical period-a time span before and after which weed competition does not reduce the crop yield, for each crop, so that weed management could be more meaningful, efficacious and cost effective. All crops need a weed-free period for about one fourth to one third of their life cycle.

Most herbicides are crop specific. It is difficult to find the most efficacious herbicides for different crop combinations. Butachlor and alachlor are such herbicides which gave satisfactory control of weeds in maize / mungbean, maize/ cowpea, sorghum / cowpea, sorghum / blackgram, intercrop systems. In maize / groundnut intercrop system, trifluralin was found quite effective. The problem of weed management in sequential crop systems is different from that of intercropping systems. Residual toxicity of herbicides applied to the previous crop assumes importance and should be considered while suggesting weed management measures.

Critical periods of competition and percentage yield reduction

Crop	Critical period (DAS)	Reduction in yield (%)	Remarks
Upland rice	10-40	60-80	Approximate % losses caused by
Maize	15-40	40-60	weeds. However the extent of
Sorghum & millets	15-35	30-40	reduction in yield mainly depends on the types of weeds present and
Cotton	30-60	45-85	other factors
Groundnut	25-55	55-70 (Erect) 20-25 (Spreading)	
Sunflower & safflower	30-60	30-50	
Castor	30-60	35-40	
Sesamum	15-30	45-50	
Soybean	20-45	25-35	
Mungbean	20-45	50-55	
Pigeonpea	25-60	40-60	
Black gram	25-35	45-50	

Off-season tillage (summer tillage) keeps the soil free of weeds, besides conserving more moisture. Crop rotations with dissimilar life cycles or cultural conditions are potential means of controlling weeds under most of the dryland situations. Inclusion of legumes, particularly fodder legumes, has proved very effective in controlling weeds.

Fertility management

In dryland areas, soil fertility management is as important as soil and moisture conservation measures. In order to get the best out of limited moisture, it is essential that nutrient requirements of dryland crops are adequately met. Highest moisture use efficiency is obtained when the crop is sown in time, has optimum plant density, and is free from weeds, diseases and pests and above all, is nutritionally well managed. Response of crops to nutrients is dependent on soil moisture status and soil type, a fact well depicted by data set out in table of the above chapter, respectively. Fertilizer recommendations, based on average annual rainfall and soil moisture storage for principal rainfed rainy and post rainy season crops are location specific. In case of rape seed-mustard, advancing the time of application of fertilizers has shown striking advantage during two out of three years.

Preparation of appropriate crop plan for dry region and mid season aberrant weather conditions

The appropriate crop plan and practices, which have so far been generated and recommended for achieving. The objective of increased and stable crop production in dry land areas have been summarized in the following points.

1. Proper and timely tillage

Primary tillage is done in order to increase water intake and minimize weed infestation. Farmers do it by country plough. It has to be carried out as early as possible, prior to the sowing seasons. Keeping this in view, the concept of off-season tillage has been developed. The practice increases soil moisture intake and results in efficient weed control. Another important advantages of this practice is better distribution of draft power requirement sand rapid coverage of the area with a given source of draft power.

Shallow off-season tillage with pre-monsoon showers ensures better moisture conservation and lesser weed intensity. It has resulted in 20% yield increase of sorghum in Andhra Pradesh. In light soils, surface crust formation reduces seedling emergence and rain water intake. These by leading to more surface runoff. A light cultivation would be adequate to open such soils. Areas with

Management of Rainfed Crops 53

hard pans, *e.g.* alluvail soils of north which tend to develop hard pans due to consistent cultivation and areas with textural profiles *i.e.* with more day as depth increases *e.g.* semi-arid red soils of southern India and sub-humid red soils of north eastern India, requires the deep tillage. Deep tillage helps in increasing water in soils having textural profiles and hard pan. This has resulted in 10% yield increase in sorghum and 9% yield increase in castor. For *in-situ* moisture conservation, land has to be opened so that it can cause hurdle to flow of rain water. Tillage machines of appropriate size and type matching the power sources need to be used.

2. Early sowing

Dry land crops have to be planted at the earliest opportunity after the onset of monsoon. Under some situations, dry sowing is desirable. Early planting helps in better crop yield by providing better moisture conservation, good seedling vigour, availability of longer growing season and avoidance of pests and diseases.

Efficient crops and their sowing time especially for Hisar region

Kharif crops	Varieties	Sowing time
Bajra	HHB-50, HHB-60, HHB-67, HHB-68, HHB-94, HHB-117, HHB-197, HHB-216	End June to 3rd week July (Direct seedling 4th week July to 2nd week August (Transplanting of 21 days old seeds)
Clusterbean	HG-365, HG-75, HG-563	2nd fortnight July but can be sown upto 2nd week of August End of June to 3rd week July
Cowpea	HFC-42-1, Charadi,CS-88	
Moong bean	S-9, Muskan	For low rainfall areas.
Moth bean	Jawala, Jadia, Marumoth, RMO-40	For normal 1st week July for late sown 1st week of August.
Rabi crops		
Chickpea	G-24 for light soils C-235 for blight susceptible areas HC-1	1st fortnight of October
Mustard	Varuna and RH-30, Luxmi, RH-819, RH-781, RB-50, RB-9901	Throughout October
Rapeseed	BSH-1	End September to 1st week of October
Taramira	T-27	Throughout October but can be upto November 15
Barley	BH-902, BH-75, BH-393	

3. Choice of suitable crops

Selecting suitable crops and varieties capable of maturing within actual rainfall periods will not only help in enhancing production of a single crop but also in intensifying the cropping intensity. The capacity to produce a fairly good yield under limited soil moisture conditions is the most desirable criteria for selecting a crop variety for dry lands. In other worlds, crop varieties for dry land areas should be of short duration, through resistant tolerant and high yielding which can be harvested within rainfall periods and have sufficient residual moisture in soil profile for post-monsoon cropping.

The depth of soil, available profile moisture and amount and distribution of rainfall are some of the important factors playing an important role in the choice of crops. The traditional and alternative efficient crops identified for different regions are listed below:

4. Plant population

The optimum plant density of a crop is related to the soil moisture condition. The important aspect of plant population is row spacing. Adoption of wider rows will help in early coverage of area of less energy cost, besides better use of stored soil moisture.

Seedling rates and planting pattern especially for Hisar region

Crop	Seed rate (kg ha⁻¹)	Row spacing (cm)		Remarks
		Inter	Intra	
Bajra	4	45	10	(60 x 10 cm)
Clusterbean	25	45	10	(30 x 15 cm) if late sown in August
Greengram	20	45	10	(30 x 10 cm) if late sown in August)
Cowpea	30	045	10	
Chickpea	50	45	—	(60 cm under low moisture conditions)
Mustard	5	45	10	(60 x 15 cm under low moisture conditions)
Rapeseed	5	45	10	(30 cm in good rainfall year)
Taramira	5	45	10	(60 cm in low rainfall year)
Barley	100	30	—	

5. Cropping systems

The rainfed areas are generally mono-cropped having a cropping intensity of 100. Farmers either raise crop in monsoon season and leave the field fallow in postmonsoon season or conserve rainwater in monsoon and take a crop in post monsoon season on conserved moisture. However, the researchers suggested that the cropping intensity could be increased through intercropping or double cropping depending upon the weather situation. Taking the rainfall, soil type and other meteorological parameters into consideration, effective growing seasons have been worked out for different seasons. Based on this information, potential cropping systems have been suggested for the country.

Potential cropping systems in relation to rainfall and soil type

Rainfall (mm)	Soil type	Effective growing season (weeks)	Cropping system
320-600	Red and shallow black soils	20	Single *kharif* cropping
350-600	Deep sierozems and alluviums	20	Single cropping in either *kharif* or *rabi*
350-600	Deep black soils	20	Single *rabi* cropping
600-750	Red, black and alluvial soil	20-30	Intercropping
750-900	Alluviums, deep black, deep red and submountaneous soils	>30	Double cropping with monitoring
>900	Alluviums, deep black, deep red and submountaneous soils	<30	Double cropping assured

5.1 Double cropping

Double cropping is possible in the areas receiving more than 750 mm rainfall with a soil-moisture storage capacity of more than 20 cm of available water for a successful double cropping, early planting of the *kharif* crop and its harvest as early as possible helps in quick establishment of the *rabi* crop in areas below Vindyan region (Malwa and Vidarbha regions) as temperature transition is a continuum and suggested crops are not too thermo sensitive. But in the northern belt because of thermal sensitivity most of the *rabi* crops cannot be planted earlier. In the process, the surface soil starts drying up after the harvest of *kharif* crop and seeding of subsequent *rabi* crop sometimes poses problem. In such situation, pre-sowing, irrigation for promoting seedling emergence might be important. Use of improved seeding devices like ridger seeder developed at Hisar center might be useful.

56   Rainfed Agriculture

Double cropping systems suggested for different regions in the country

Region	Crops
Sub mountain north-west region	Rice-wheat
	Rice-chickpea
	Maize-chickpea
	Soybean-wheat
Bundelkhand region of M.P.	Rice-wheat
	Rice-chickpea
	Sorghum-chickpea
	Greengram-wheat
Plains of Eastern U.P.	Rice-chickpea
Sub humid soils of Chotanagpur	Rice-linseed
	Maize -safflower
Sub-mountain soils of North – East Punjab	Maize-mustard
	Maize-chickpea
Malwa plateau of M.P.	Maize-safflower
	Sorghum-safflower
	Sorghum-chickpea
	Soybean-safflower
Bundelkhand region of U.P.	Cowpea – mustard
	Sorghum – chickpea
Hisar region	Cluster bean – taramira
	Clusterbean – gram

5.2 Intercropping

In the seventies, agriculture became more production oriented after the availability of high yielding, short duration and inputs responsive crop cultivars. This changed the concept of intercropping to produce more over space and time. Keeping this in view, research efforts have been made to identify the most remunerative intercropping systems for different regions in the country. While doing so due importance is given to most common crop of the region and to pulses and oil seeds. The improved intercropping systems recommended are given below :

Management of Rainfed Crops 57

Improved intercropping systems suggested for different regions in the country

Region	System	Ratio
Vidarbha region of Maharashtra	Sorghum+green gram	1:12
	Sorghum + pigeonpea	
Malwa plateau of M.P.	Sorghum+ pigeonpea	4:1
Black soils of South-east Rajasthan	Sorghum+green gram	1:11
	Sorghum + pigeonpea	
Deccan region of Maharashtra	Pearl millet + pigeonpea	4:1
Medium black soils of Rajkot region	Pearl millet + pigeonpea	4:16
	Groundnut + pigeonpea	
Sub-humid red soils region of Orissa	Rice + pigeonpea	8:2

6. Timely weed control

Off season tillage, proper seed bed preparation and timely sowing helps the crop in combating weeds. Studies conducted in the project very clearly showed that the first 25 days period in case of 100 days crops and 35 days period in case of 150 days crops is sensitive to weeds. Therefore, one or two weedings with blade harrow or country plough keeps the weeds well in control. The effects are still better when the practice of inter-culture operation with bullock drawn implements along with hand pulling of intra-row weeds is adopted and thus increase crop yields appreciably. For this purpose some implements like blade harrow, rotating weeders like dryland weeder are recommended. Though the herbicides are quite effective in controlling weeds. The high cost of these chemicals takes them beyond the reach of dryland farmers.

7. Use of fertilizers

Soils of drylands in the country are not only thirsty but hungry also because these soils are severely eroded horizontally as well as vertically whenever efforts are made towards bunding and leveling of fields in dry land areas, it is the surface soil that is removed. The resultant effect is that the fields are rendered shallow in depth and completely deprived of plant nutrients, particularly nitrogen, phosphorus and potassium. It is therefore, necessary to apply all the three major nutrients in adequate amounts. Fertilizer use plays in drylands. Not only that, its use insulates crops against moderate moisture stress as the root system would be more developed compared to unfertilized crops. Thus fertilizer use helps in more efficient use of moisture. Since soil moisture is limiting in drylands, the availability of nutrients becomes limited, attempt should always to made to apply fertilizers in furrows below the seed if seed cum fertilizer drills drawn by bullock or tractor are available. This very objectives can be fulfilled. There has been belief among the farmers of

58　Rainfed Agriculture

dryland areas that use of fertilizer increases the chances of crop failure, but recent findings have shown that the use of fertilizer is not only helped in providing nutrients to crop but also helpful in efficient use of profile soil moisture. In dry land areas, a proper mixing of organic and inorganic would be desirable, organic forms have row nutrient content, but help to improve the moisture holding capacity of soils. In addition to yield advantage, nutrients, like potassium, help to increase drought tolerance by affecting plant-soil-water relationship. Transpiration losses are reduced and productivity per unit water increases.

For the maximum benefit from the applied fertilizer there is need for maximum fertilizer use efficiency that can be achieved by proper placement and split application of fertilizers.

Fertilizer recommendation for dry region especially for Hisar region.

Crop	Nutrients (kg/ha)		Method of application	Remarks
	N	P_2O_5		
Bajra	40	20	In 2 splits, ½ drilled at sowing + ½ at knee high stage	Apply depending on soil test
Clusterbean	0	30	Placement	
Green gram	20	40	All basal drilled	
Chickpea	20	40	Drilled	
Mustard	40	20	Drilled	Adopt advance placement of
Barley	40	20		fertilizer at the end of monsoon season

40 kg P_2O_5/ha to chickpea if grown after bajra. If sown after *kharif* fallowing adjusts P_2O_5 based on soil test.

Crop-wise specific impact points and recommendation

Bajra

I. Plant stand

The following practices help to establish a satisfactory plant stand

Wider row spacing of 45 cm instead existing traditional practice of closer row spacing of 30 cm is recommended. In wider row spacing the seed may not be buried too deep due to coverage of soil from the adjacent furrow. Wider spacing also allows interculture and weeding operations with improved implements.

Sowing of the crop by a plough with narrow bottom or cotton seed drill of 2-bottom-tractor drawn ridger seeder. In case of 2-bottom tractor drawn ridger seeder the paired row spacing of 30 cm and between two pairs of rows 90 cm is recommended.

Gap filling by transplanting of seedling. The seedlings may be obtained either by raising nursery in separate plots or from the thick stand in the same field of adjacent field.

II. Weed control and moisture conservation

The practices that control weeds also conserve moisture. The following simple practices are recommended for timely control of weeds and conservation of moisture

1. Bunding of field

Most of the fields in dryland areas are open without any bunds. At places there are ups and downs with undulating topography. Putting of at least strong boundary bunds-around each acre of the field and land shaping of ups and downs could go a long way to conserve the precious and scarce amount of rainfall water. Practice of bunding may be adopted by the farmers in their cultivated fields where there is no problem of shifting of sand dunes. Agro-forestry programme as advocated by Department of Forest should be extended to stabilize the sand dunes and overcome the problem of wind erosion.

2. Preparation of weed free seedbed

Whenever possible the field should be harrowed well in advance before sowing of the crop when the land fallow during summer or winter months. Post harvest tillage just after harvesting of the previous crop is very useful to reduce intensity of weeds in the following season. Use of blade harrow 'randha' and triphali is very useful in light texture soils as a primary tillage implement or even in medium texture soils as a secondary tillage implement.

Preparation of a weed free seedbed well in advance before sowing of bajra crop also helps in timely planting of crops as soon as rains are received.

In dunal areas where there is a problem of wind erosion, the weeds may be cut by rantha during summer so as to provide a straw mulch of the weeds instead of allowing the seeds to grow and to sprout in the soil, limited supply of moisture and nutrients.

3. Timely weeding and interculture operations with improved implements

Use of hand wheel, hoe, bullock or camel drawn blade hoe not only promote efficient and timely control of weeds with 1/3 reduced cost of labour, but also helps better conservation of soil moisture by increasing rate of infiltration and reducing loss of water due to evaporation.

Weeding could be started with these implements after 2 to 3 weeks of sowing of the crop when the weeds are young and easily controllable. Interculture operations should be done after every effective dower to break the surface curst and to increase infiltration rate of rainwater.

4. Inter row water harvesting *in-situ*

Making of ridges and furrows with the help of ridger either at the time of planting of the crop or during interculture operations in standing crop helps in harvesting and conservation of moisture *in-situ*. This practice also mitigates adverse effect of sometimes heavy down pour of rainfall by providing adequate drainage facility. In sloppy fields sowing of the crop should be done across the slope and preferably along contours.

5. Fertility management

In dryland areas the soil fertility could be improved by raising good legume crops (Guar, moong and gram) in rotation and by supplemental use of fertilizers.

Application of 40 kg nitrogen and 20 kg P_2O_5 per year and 25 kg $ZnSO_4$ ha (once in three years) in deficient soils on the basis of soil test is adequate for obtaining good yield of bajra crop when it has been rotated with a high yielding legume crop of gram or guar. The basal dose of fertilizers should be drilled by local plough at the time of sowing. Half of nitrogen is to be applied as basal and other half as top dressing in standing crop at the time of final interculture and weeding operations, when soil moisture condition is favourable.

For getting maximum response and returns from the use of fertilizers, proper plant stand, maintenance of weed free condition and adoption of moisture conservation practices are must.

Guar and mung

Application of 20 kg N and 40 kg P_2O_5 and 25 kg $ZnSO_4$/ha (once in three years) in deficient soils based on soil test serves as booster, for obtaining their satisfactory yields.

Crop	Variety
Pearl millet	HHB-50, 60, 67,94,117,197,216.
Guar	HG-75, HG-365, HG-563
Moong	S-9, Muskan
Cowpea	Chirodi (grain), HFC-42-1 (fodder), CS 88
Arhar	UPAS-120, Manak
Maize	Ganga-5, Vijay composite
Groundnut	Pb Moongphali No. 1
Urd	T-9

Seed rate and spacing

Crop	Seed (kg/ha)	Plant-plant spacing (cm)
Pearlmillet	3.75-4.0	45cm, 30:60 paired
Guar	17.5-20	45
Moong	15-20	45
Cowpea	15-20	45
Arhar	125-15	45
Groundnut		
PM-1	87.5	30 x 22.5
M-145	112.5	30 x 15
MH-2	150	15 x 15
Sorghum + cowpea	50+10	25
Urd	20	30

Special package of practices of *rabi* crops in dryland areas of Haryana

$1/4^{th}$ of cropped area in Haryana is rainfed. We can divide Haryana into two parts based on this major dryland area (87%) in South-west zone adjoining Rajasthan which includes Hisar, Bhiwani, Mahendergarh, Rewari, Gurgaon, Jhajjar and parts of Jind. Average rainfall in this area is 250-500 mm of which 80-85% occurs in monsoon period. The soil is light to medium textured having sand dunes in some area and underground concrete layer. WHC is low in major crops like gram, raya and taramira. In Mahendergarh, Gurgaon and Rewari barley is also taken. For taking *rabi* crops people leave *kharif* follow and conserve monsoon rains. If rains occur in September farmers take gram after pearl millet. If rains occur in 1st fortnight of December than late taramira can also be taken. Better moisture conservation is the leaf for successful dryland crops.

In North-East zone of Haryana comprising Ambala, Yamunanagar, Panchkula, parts of Sonepat, Gurgaon, Faridabad and Rohtak where annual average rainfall is 500-1000 mm of which 75-80% occurs from July-Sept. Rainfed crops are taken in Shivalik foot hills. Due to undulating topography and soil erosion, gullies are formed and due to poor moisture conservation the soils are less productive. The soils a sandy loam to clay loam having low organic matter hence WHC is low. Major crops are wheat, raya, toria, gram and lentil moisture conservation and its controlled use in by factor.

A) *Soil and water conservation:* To conserve more and more rain water the fields should be levelled and well bunded, bunds should be made across the slope to conserve more water. Bunds should be atleast 45 cm high. Contour

62 Rainfed Agriculture

cropping should be practiced on slopes. By using antitranspirants, water loss through evaporation can be minimized along with better weed control.

B) *Rain water harvesting*: During rainy season due to heavy rains some times soil cannot conserve entire rain water, if it is not controlled properly it can cause soil erosion etc. This rainwater can be harvested in the field at the lower side of the slope in a 2-3 ditch having 1:1 slope in boundaries. This harvested water can be used in pre-sowing irrigation of rabi crops.

C) *Field preparation*: In these areas shallow ploughing is done by harrow or cultivator but in clay loam soils deep ploughing is done once in three years. If moisture is less in the top layers than heavy planting or heavy rolling is done.

D) *Selection of crops*: In heavy rainfall areas (Ambala, Y. Nagar) wheat, raya and toria and in low rainfall areas raya and gram are profitable crops. Raya and toria is taken generally after fallow in S-W monsoon and gram after fallow in N-E monsoon. Crops are selected based on moisture availability in the field. For *e.g.*

Crop	Moisture availability
Taramira	100-125 mm / mt
Raya and gram	125-175 mm / mt
Barley	>175 mm / mt

Intercropping of raya in gram : Raya can be intercropping in gram in the ratio of 6:1 or 8:1 and 2.5 t/ha yield of raya can be taken this way.

E) *Choice of varieties*

Crop	Variety
Wheat	C-306, WH-147, WH1025.
Raya	RH-30, Varuna, RH-781, RH-819, RB 24, RB 50, RB 9901 (Geeta)
Toria	Sangam, TH-68, TL-15
Gram	C-235, Gaurav, Haryana Channa No. 1
Lentil	L 9-12
Barley	BH 902, B.H.-393, B.H. 75
Taramira	T-27

F) *Sowing time*: Raya is sown in second to 3rd week of October, gram and barley from last week of October to 1st week of November while taramira can be taken upto last week of November Toria should be sown upto mid September and wheat in entire month of November.

Management of Rainfed Crops 63

G) *Seed rate and method of sowing*: For gram, use 35-45 kg/ha, raya and taramira 5 kg/ha, barley 75-80 kg/ha, wheat 100 kg/ha and lentil 30-35 kg/ha. seed of gram, raya and taramira should be sown at 45 cm row to row distance, wheat and barley at 25 cm row to row distance. In the years of very low soil moisture, in the months of December, January if plants show moisture stress symptoms, 25% plants can be removed, sowing is done with ridger seeder. Gram can be sown at 10 and 12 cm depth by pora method using desi plough. Barley and wheat is sown by seed drill. In case of very low moisture at sowing time, make furrows by ridger seeder and sowing is done by desi plough in these furrows for gram and raya.

H) *Fertilizer use*:

Crop	Fertilizer (kg/ha)	
	Nitrogen	Phosphorus
Wheat (C-306)	30	15
Wheat (WH-147)	60	30
Gram	20	40
Raya / Toria	40	20
Taramira	20	---
Barley	30	15
Lentil	20	40

In potassium deficient soils apply 15 kg K_2O/ha in wheat.

Apply entire fertilizer at the time of sowing : 10-15% higher yield of raya can be taken if fertilizer is applied at the time of sowing.

I) *Inter-culture*: Weeds must be removed in time. Weeding can be done by wheel hand hoe which along with weeding conserves moisture also.

J) *Alternate land use system*: In soils not capable of raising crops and have capacity to conserve moisture, Ber and Janti can be taken. Agro-forestry and agro-horticulture can be profitable practiced by taking up Anjan grass and pulses for first 4-5 years.

Hints for higher productivity

1. Better rain water conservation.

2. Well seed bed preparation.

3. Selection of crops according to moisture availability.

4. For better germination proper sowing with optimum seed rate.

5. Plant stand adjustment according to moisture.

6. Improved equipment interculture.

7. Weeding at proper time.

8. Balanced fertilize use.

9. Control of insect pest and diseases at right time.

10. Proper storage of grains and proper time marketing.

Crop planning and substitution in Dry region

In India, nearly 70 per cent food grains are grown in dryland farming areas. However, the output is less than proportionate – 42 per cent of the total food grains output (Srivastava, 1987). Area wise, drylands constitute 68 per cent of total cultivated area in our country. Dry farming areas or regions are characterized by spatial and temporal variations in rainfall, in-fertile and relatively shallow soils and a growing season varying from 100 to 200 days. Dry land tilt the balance of food production in the country. Droughts are common, yields have been low and highly unstable. The problems of crop production are many and complex. Stabilization of crop yields at a respectable level seems to be the rational approach for dry land development.

There are various ways and means to increase and stabilize production on dryland. Increasing cropping intensity through proper crop planning and total production by the substitution of in-efficient crop with efficient crops in that region are of paramount importance among the other suggested alternatives.

Crop planning

Crop planning refers to systematic arrangement of various crops to be grown in a particular region, taking into consideration the available resources, environmental factors, managerial ability of the growers and availability of production technology with a view to maximizing production per unit area. The farmers aiming at higher profits should resort to crop planning based on sound, scientific principles.

The main objectives of crop plan area

1. To increase farm income by better use of available resources and technical know-how.

2. To reduce the costs by saving on unnecessary and uneconomic expenditure and

3. To set forth a programme of tasks to be done.

Management of Rainfed Crops 65

4. There can be no single perfect cropping plan suitable for adoption by farmers in a region. Every individual farmer has to have a plan to suit his conditions.

The factors to be considered before the preparation a cropping plan are:

1. Soil type.
2. Climatic conditions.
3. Cost, income and risk involved in growing a particular crop.
4. Capital, labour, other resources and facilities available.
5. Scientific advances made and
6. Special needs of the farmers

The crop planning and formulation of cropping pattern in an area depends mainly on two factors, *viz.*, soil and climate. Climate has special significance in dry regions, since the availability of soil moisture during the cropping season over rides any other consideration in crop planning. The choice of crops and suitable varieties for a given area in a given cropping season will depend upon the variations in precipitation received in that area and potential evapotranspiration. A study of relative magnitudes of water surplus and water deficiency in different dry farming zones of the country is of paramount importance to crop planners. It helps them to identify suitable crops and cropping patterns according to water availability during the crop season. From water balance studies, one could work out the assured of safe periods of crop growth and thus structure the cropping patterns accordingly. The safe cropping season could be worked out based on moisture index (M.I.) which is calculated as:

$$\text{Moisture index (M.I.)} = \frac{P - PE}{PE} \times 100$$

Where, P = Monthly precipitation (mm)

PE = Monthly potential evaporation (mm)

On the basis of moisture index, Indian Council of Agricultural Research (1970) has divided the dry farming tracts of the country into four zones.

Zone	Group	M.I.	Region
I	Arid	More than 66.7	Jodhpur, Hisar, Rajkot, Bellary and Anantpur
II	Semi-arid	-66.7 to -33.3	Udaipur, Ludhiana, Delhi, Agra, Jhansi, Indore, Solapur, Akola, Hyderabad, Bijapur, Kovilpatti, Anand and Hebbal
III	Dry sub-humid	-33.3 to 0.0	Varanasi, Rewa and Samba
IV	Moist sub-humid	0 to + 20	Bhubaneswar and Ranchi

Dehradun is a special problem station with a moisture index of +86. Similarly, soil zones in dry farming tracts have also been identified for different stations (Dryland Workshop, 1972).

Conclusion

Crop production in the dry regions of India can be increased and stabilized by resorting to scientific crop planning and crop substitution. Soil and climatic factors particularly moisture availability in crop season along with the availability of needed resources and proven production technology are to be considered for successful crop planning. Possibilities of increasing cropping intensity through inter and sequential cropping systems in different agro climatic zones of the dry farming tracts in India have been explored. Similarly, the scope of substitution of traditional, in efficient crops with non-traditional and efficient crops in different zones have also been discussed based on the results of several experiments conducted in different parts of the country under "All India Coordinated Research Project for Dryland Agriculture".

4

Water Harvesting and Moisture Conservation

Natural resources conservation and their management hold key to sustainable agriculture and livestock production. It is all the more crucial for countries with predominant agrarian economies where development of sustainable agriculture is essential for overall growth, redressal of poverty and security. Conservation of both soil and rain water as very crucial and basic resources have been practiced since ancient times in India. However, there has been renewed emphasis on conservation and efficient utilization of these resources in the recent past. The fact that 70% of the arable area of India is rainfed with precarious supply of water and that rainfed agriculture supports 44% of India's human population and contributes 90 per cent of coarse cereals and pulses, 80 per cent of oilseeds and 65 per cent of cotton and growing realization that further gains in productivity of crops and livestock will emanate from rainfed regions, leave no room for complacency in this regard.

Year after year, the fate of a vast majority of Indian farmers hangs in balance, as success with rainfed agriculture continues to be a gamble. It is evident that crop yields in semi-arid areas depended more on rainfall distribution than on total rainfall and lack of serious efforts to create the water supply for crops through scientific management of rainwater is a factor favoring this avoidable uncertainty. In rainfed agriculture no other input can perhaps enhance the yield without effectively tackling of the rainfall aberration related sub-optimal moisture availability. Therefore a prerequisite for substantial improvement in the agriculture production in the semi-arid region is to manage runoff water and to use it either at the time of moisture stress even during the monsoon or in next season. It has been reported that only one supplementary irrigation at proper stage can double of *rabi* crops. But supplemental water is a developed resource and is more expensive than the natural resources. Hence, it is all the more necessary to use this water judiciously and efficiently. Standardization of techniques by which as much of precipitation as possible can be conserved for crop use, either directly in the soil profile through infiltration or through runoff collection and recycling is an area of priority in research.

Rainwater availability in India

On an average, India receives an annual rainfall of 1,100 mm, which ranges between 100 mm in Western India (Thar desert) to over 10,000 mm in the northeastern parts. India can be divided into four major zones based on the average rainfall.

Measured in terms of irrigation water equivalence, India annually receives 400 million hectare meters (M ha m) of rainwater. India's share thus works out to about 4% of the global precipitation (11 billion ha m). In fact, India receives the largest amount of rainwater among the countries comparable to land size. The southwest monsoon (June-September rains) contributes 74% of the total precipitation. In contrast, the northeast monsoon provides 30%. The remaining 23% is supplied during pre- (~10%) or post-monsoon (~13%) seasons.

Out of 400 M ha m rain water being received in the country, 178 M ha is available as surface flows. Big reservoirs can at best store about 70 M ha m. Nearly 24 M ha m, in surface flow is possible to stash at the primary source or the donor areas. Approximately 6.32 M ha m is harvestable through field level structures in low to medium rainfall regions (~1000 mm and less), where certainty in agricultural productivity is more often uncertain.

Water harvesting and recycling

It involves collecting excess runoff, natural or induced, in surface reservoir for agricultural use. Water thus harvested can be stored in surface reservoirs or ponds or in soil. Appropriate water harvesting strategies depend on soil type, terrain, and rainfall characteristic.

Water harvesting

'Water harvesting' is generally used as an umbrella term covering a range of methods of collecting and conserving various forms of runoff from different sources. Water harvesting can be done both *'in situ'* and dugouts /ponds depending upon the situation as well as socio-economic status/capabilities of the farmer. In dryland agriculture, water harvesting usually denotes the collection of excess runoff in a storage tank and using it for the battement of crop production in the collected or other areas.

The water collection in the farm pond is directly used for protective irrigation. The water stored in other structures will recharge the ground water and used for protective or supplementary irrigation. Collecting and storing water on the surface of the soil for subsequent use is known as water harvesting. It is a method to induce, collect, store and, conserve local surface runoff for agriculture in arid and semiarid regions. Water harvesting is done both in arid and semiarid regions with certain differences. In arid regions, the collecting area or catchments area is substantially in higher proportion compared to command area. Actually, the runoff is induced in catchment area in and lands, whereas in semiarid regions, runoff is not induced in catchment area, only the excess rainfall is collected and stored. However, several methods of water harvesting are used both in arid and semiarid regions.

Runoff from cultivated fields should be led through grassed waterways to suitably located excavated or embankment farm ponds. The runoff harvesting, storage and utilization involves -

(a) Selection of catchment;

(b) Selection of site and estimation of size of farm pond;

(c) Construction of farm pond including inlet;

(d) Laying of silt trap and spillways;

(e) Minimization of seepage, percolation and evaporation losses;

Utilization of stored water for life saving irrigation of *kharif* crops or for establishment of *rabi* crops in the catchment.

A) Water harvesting *in situ*

Water harvesting techniques adopted *'in-situ'* are inexpensive to augment rain water use efficiency in drylands, any process of reducing the runoff volume will increase moisture infiltration in the soil leading to increased moisture volume. *'In-situ'* water harvesting and runoff recycling are the potent measures of crop life saving during periods of moisture stress. The method of *'in-situ'* water harvesting helps in improving the crop stand, especially on small seeded crops like pearl millet by mitigating the adverse crust effects on germination and seedling burying problems due to rain during germination. This practice also helps in safe disposal of excess rainwater in high rainfall areas. Appropriate land configuration for *'in-situ'* moisture conservation are broad-beds and furrows, graded border, strips, inter-row and inter-plot water harvesting systems. For shallow alfisols and related soils even dead furrows serve this purpose very well. Strategies for *'in-situ'* moisture conservation involve contour farming and installing barriers on the contour.

a. Contour farming

Contouring or contour farming implies performing all farm operations on the contour rather than up and down the slope. Contour farming is one of the easiest and most effective and low cost method of controlling erosion and conserving moisture. With contour farming, tillage operations are carried out along contours. It creates numerous ridges and furrows, which increase surface detention capacity, increase time for water to infiltrate, and decrease runoff amount, harvesting sizable amount of runoff inside the field itself. Contour farming is however, effective on gentle slopes of up to 5%. Steeper slopes require slope management techniques. Contour farming can be sometimes inconvenient because it possibly involves frequent turning of farm vehicles and loss of area, which has to be put into buffer strips.

b. Strip cropping

Another aspect of contour farming is strip cropping or growing crops in long narrow strips established on the contour. In this system, open row crops (*e.g.* corn) are grown in alternate strips with close canopy crops (*e.g.*, soybean, alfalfa, *etc.*). The close canopy crop is often grown in a contour strip down slope from the open row crop. It is also important to establish buffer strips on the contour. There are various types of strip cropping. On the basis of objectives, vegetative materials used, and field design adopted, buffer strips are called contour strip cropping, buffer strip cropping, barrier strips, border strips, or field strips. In addition to reducing runoff amount, establishing strips against the prevailing wind direction also decreases wind erosion.

c. Live beds

One or two live beds (2-3 m wide) can be created either on contour or on grade in the inter bund space. The vegetation on the beds may be according to the liking of cultivator; it can be annual or perennial or a combination of both.

d. Vegetative hedges

Vegetative hedges are established on the contour to create a barrier and increase time for water to soak into the soil. Vegetative hedges are mostly established from bench-type grasses. A widely adapted grass for tropical eco-regions for vegetative hedges is Vetiver *(Vetiveria zizanoides)* or khus grass. Vetiver is densely tufted bunch grass, which can be easily established. In addition to grasses, vegetative hedges can also be established from woody perennials, *e.g., Leucaena leucocephala, Gliricidia sepium,* etc. Properly established and adequately maintained. Vegetative hedges decrease runoff velocity, encourage sedimentation, and reduce runoff and soil erosion. However, closely spaced and narrow strips of grass or woody perennials are likely to be more effective in reducing runoff and soil erosion than narrow or single-row hedges.

e. Tied ridging

This is a modified version of ridge-furrow system where ridges are "tied" at 15 to 20 m interval to allow rainwater collection in the furrows and its infiltration into the root zone. The system not only conserves rainwater but also substantially reduces sediment and nutrient losses induced by runoff.

f. Broad bed and furrows

Broad bed and furrow system implies shaping alternate bed and furrows. This technique is especially suited to black soils, where crops are sown on preformed beds. This system is made before the season and is maintained year after year. The planting is done on the beds. Generally, the depth of each furrow is kept 0.15 m and the inter furrow spacing is maintained at 1.5 m.

g. Dead furrows

Dead furrows are laid across the land slope in rolling lands, to intercept the runoff. The spacing between dead furrows varies between 2 to 5 m or 4 to 7 crop rows. This system works well in alfisols.

h. Bunds

These are low height earthen embankments constructed across the slope in cultivated lands after deciding location of waterways. The bunds function to intercepting runoff, increase infiltration opportunity time and dispose excess rainfall safely. Such bunds can be either contour bunds or graded bunds; many-a- time, these bunds are adjusted with field boundaries, if deviation from grade or contour is not too much; no doubt, spacing depends on many engineering considerations.

i. Contour bunds

These are recommended in dry farming areas with light textured soils of slopes up to 6 per cent and where annual rainfall does not exceed 600 mm. They are designed for an expected runoff of 24 hours duration and 10 years frequency. Surplus arrangements (waste weirs) are provided to dispose of the excess runoff beyond the design storage. The cross section area of contour bunds follows the depth and type of soil; however, 0.5 m^2 is minimum.

72   Rainfed Agriculture

A small section bund may be created across the slope at half of the vertical bund spacing. Such bunds be nearly 0.1 m² in section; may be renovated at an interval of 2 to 3 years.

j. Graded bunds

These are constructed in medium to high rainfall areas in permeable soils (vertisols), having 2 to 6 per cent slope. They are also quite suitable for the soils having crust formation tendency like red 'chalka' soils of Telengana region of Andhra Pradesh. By and large; graded bunds with 0.3 to 0.5 m² section are constructed with longitudinal gradient of 0.2 to 0.4 per cent depending on the site condition. Graded bunds particularly are made with the following objectives:

(i)   To reduce runoff

(ii) To reduce soil loss; and

(iii) To divert runoff

Thus, graded bunds along with waterways and water harvesting structures not only check soil erosion, but they also provide an ideal rainwater management system for many watershed situations.

k. Compartmental bunds

Compartmental bunds, converting the area into square/rectangular plots are useful for temporary impounding of water for improving moisture status of the soil. These are made using bund formers. In medium deep black soil, they are found advantageous in storing the rainfall received during the rainy season in the soil profile there by augmenting the soil moisture for use by rabi crops. The size of the compartments may be fixed.

Compartmental bunds are usually practiced in deep black soils with relatively flat terrain and under low to moderate rainfall. The field is laid out into compartments of 6m x 6m to 10 m x 10 m using a bund former. The harvested water in these compartments conveniently infiltrates into the root zone and conserved in situ.

l. Terracing

Cultivated lands having land slopes above 10 per cent particularly in hilly areas - should be put under bench terracing by converting the lands into series of platforms. The width of bench terrace depends on the land slope and the permissible depth of cut. Bench terracing is very much effective in reducing soil erosion in hilly areas. At Ootacamund, it has helped in bringing down the

Water Harvesting and Moisture Conservation 73

annual erosion rate from 39 t/ha to less than 1.0 t/ha on 25 per cent sloping lands. At places where scattered stones are available, loose stone walls can be made to act as risers for bench terrace construction.

B) Runoff harvesting

In rainfed areas, one or two occasional heavy rainfall occurs which creates run off. In vertisols, the runoff is about 60 % and in alfisol it is about 40 %. Hence, efforts should be made for proper utilization of the run off after taking measures for *'in-situ'* soil moisture conservation.

Methods of runoff harvesting

The different methods of water harvesting that are followed in arid and semiarid regions are discussed separately.

I) Arid regions

The catchment area should provide enough water to stature the crop and the type of farming practiced must make the best use of water. In general, perennial crops are suitable as they have deep root systems that can use runoff water stored deep in the soil, which is not lost through evaporation.

Runoff farming

Ancient runoff in the Negev desert in Israel had several cultivated fields fed by water from watersheds of 10 to 50 hectares. The watersheds were divided into small catchment areas of one to three hectares that allowed runoff water to be collected in easily constructed channels on the hillsides and were small enough to prevent uncontrollable amounts of water. The channels that led the water to cultivated fields were terraced and had stone spillways so that surplus water in one field could be led to lower ones. Farmers constructed small check dams with rocks across the small gullies and guided the water to fields.

Water spreading

In arid areas, the limited rainfall is received as short intense stones. Water swiftly drains into gullies and then flows towards the sea. Water, is lost to the region and floods caused by this sudden runoff can be devastating often to areas otherwise untouched by the storm. Water spreading is a simple irrigation method for use in such a situation. Flow waters are deliberately diverted from their natural courses and spread over adjacent plains. The water is diverted or retarded by ditches, dikes, small dams or brush fences. The wet flood plains or valley floods are thereby used to grow crops.

Microcatchments

A plant can grow in a region with too little rainfall for its survival if a rainwater catchment basin is built around it. Microcatchments used in the Negev desert range from 16 m². Each is surrounded by dirt wall of 15.20 cm height. At the lowest point within each microcatchment, a basin is dug about 40 cm deep and a tree is planted in it. The basin stores the runoff from microcatchment.

II) Semi-arid regions

Water harvesting techniques followed in semiarid areas are numerous and also ancient. The objectives of water harvesting structures are:

(i) To store water for supplemental and off-season irrigation,

(ii) To act as silt detention structure;

(iii) To recharge the ground water;

(iv) To raise aquaculture species; and

(v) Recreation and allied agricultural uses.

Thus in most situations, in addition to reviving traditional water harvesting systems, other measures are also needed. Some of such typical water harvesting structures adopted in watershed are as follows:

Dug wells

Hand dug wells have been used to collect and store underground water and this water is lifted for irrigation. The quality of water is generally poor due to dissolved salts.

Tanks

Runoff water from hillsides and forests is collected on the plains in tanks. The traditional tank system has following components viz. catchment area, storage tank, tank bund, sluice, spillway and command area. The runoff water from catchment area is collected and stored in storage tank on the plains with the help of a bund. To avoid the breaching of tank bund, spillways are provided at one or both the ends of the tank bund to dispose of excess water. The sluice is provided in the central area of the tank bund to allow controlled flow of water into the command area. The command area of many tank ranges from 25 to 100 hectares. In areas receiving annual rainfall of about 1000 to 1500 mm with runoff around 50 per cent, where canal and well irrigation is not feasible due to topography and underground water table, tanks are found suitable. An earthen

Water Harvesting and Moisture Conservation   75

dam of 12 inch height stores about 55,600 inch of water from a forest watershed of 10 hectares. This tank can provide supplemental irrigation for 20 hectares of rainfed farm land. However, there are a few tanks with command area of more than 1,000 hectares. Unlike wells, the quality of water is good in tanks. Water from the tanks is used to irrigate the command area by gravity flow.

Percolation tanks

Flowing rivulets or big gullies are obstructed and water is ponded. Water from the ponds percolates into the soil and raises the water table of the region. The improved water level in the wells surrounding the percolation tanks is used for supplemental irrigation.

Minor irrigation tanks

Minor irrigation tanks are constructed across the major streams with canal system for irrigation purpose by constructing low earthern dams. A narrow gorge should be preferred for making the dam in order to keep the ratio of earthwork to storage as minimum. The height of the dam may vary from 5 to 15 m. The tanks are provided with well-designed regular and emergency spillways for safety against side cutting. In micro-watersheds, water harvesting *bundhies,* similar to small scale irrigation tanks are also recommended. By and large, the water harvesting *bundhies* are not integrated with extensive canal system.

Nala bunds and percolation tank

Nala bunds and percolation tanks are located in the nalas having permeable formations with the primary objective of recharging the ground water. A strict regulation on the silt load entering the downstream reservoirs is an additional advantage of percolation tanks. The percolation tanks encourage the digging up of wells downstream of recharged area for irrigation purposes. The percolation tanks; are provided with emergency spillways for safe disposal of flow during floods.

Stop dams

Stop dams are permanent engineering structures constructed for raising the water level in the nala for the purpose of providing life saving irrigation during drought periods. These are located over flat nalas at narrow gorges carrying high discharge of long durations. They are created over stable foundation conditions where hard rocks are encountered. For the stop dam, a site with larger water storage capacity would be desirable.

Farm ponds

Farm ponds are typical water harvesting structures constructed by raising an embankment across the flow direction or by excavating a pit or a combination of both. Dug out ponds in light soil require lining of the sides as well as the bottom with suitable sealants. This is not the case with heavy black soil. Considering the benefit of the harvested water, different type of linings, viz. brick, concrete, HDPE, *etc.* may be used.

Types of farm pond

Farm ponds are small storage structures used for collecting and storing runoff water. As per the method of construction and their suitability for different topographic conditions farm ponds could be classified into three categories,

(i) Excavated farm ponds suited for flat topography,

(ii) Embankment ponds for hilly and rugged terrains with frequent wide and deep watercourses, and

(iii) Excavated-cum- embankment type ponds for soils having mild to moderate sloping topography

Excavated farm ponds are of three types according to their shape. They are:

1. Square
2. Rectangular
3. Circular ponds

Among these, the circular ponds have the geometrical advantage that they have the highest storage capacity and have least circumferential length for a given surface area and side slopes. However, their curved shape is disadvantageous in as much as substantial area is normally lost for agricultural operations. Hence, either square or rectangular ponds are suggested.

Selecting a catchment for a farm ponds

Selection of catchment for locating a farm pond is extremely important for several reasons. It is selected on the basis of its potentiality for yielding sizeable quantity of runoff. Too big catchment results in a rapid silting, while too small a catchment may not bring in enough water into the pond. The runoff amount from different catchment will depend on several factors tabulated below:

Factors governing amount of runoff

Major factor		Associated factors
1. Rainfall	1.	Rainfall intensity
	2.	Rainfall duration
	3.	Rainfall distribution
	4.	Events of rainfall causing runoff.
2. Land topography	1.	Degree of land slope
	2.	Length of run
	3.	Size and shape of the catchment
	4.	Extent of depressions and undulations on the catchment.
3. Soil type	1.	Soil infiltration rate
	2.	Antecedent soil moisture
	3.	Soil texture
	4.	Soil structure
	5.	Soil erodibility characteristics.
4. Land use pattern	1.	Cultivated or uncultivated or partially cultivated
	2.	Under pasture or forests
	3.	Bare fallow or with vegetation
	4.	Soil and moisture conservation measures adopted or not
	5.	Crop cultural practices that are adopted.

Recycling of water, harvested and stored in farm ponds

Runoff collected into farm ponds can be used in the following situations.

a) Saving a standing *kharif* crop from drought effect.

b) Providing pre-sowing irrigation for *rabi* crop.

c) Extending the growing season for the benefit of long duration crops (castor) and

d) Providing a minimal irrigation for growing vegetable/fruit or fodder trees in small areas.

The harvested water needs, to be judiciously used for crop life saving/ supplemental irrigation to increase and stabilize crop yield and to increase cropping intensity. The quantity of water to be applied as irrigation should be just enough to take care of the critical period and to cover as large an area as possible. Since the amount of water is sufficient for only one or two irrigations, it is essential to decide the time of irrigation. Supplemental irrigation can be scheduled by two approaches viz. soil moisture depletion approach and critical stage approach. Irrespective of the stage of crop, irrigation is scheduled when soil moisture approaches permanent wilting point to save the crop. This is called life-saving irrigation. Scheduling irrigation at critical stage of crop gives more return per unit area of water applied.

For crops like sorghum, flowering seems to be most critical stage. In tobacco, irrigation at 3 weeks before topping has given maximum benefit. The yield increase due to life saving irrigation varied depending upon the crop and the region.

Water stored in farm ponds is generally a small quantity due to limited catchment area and management practices adopted. The usual amount of water given per irrigation is 5 cm under irrigated conditions. As the amount available is less under dryland conditions and large areas are to be irrigated, a minimal irrigation method is adopted. Irrigation water is applied through a furrow 10 cm wide and 20 cm deep and 10 cm long. The amount of water applied in this method ranges from 10 to 20 mm per irrigation to 20 mm per irrigation.

In-situ moisture conservation

The second approach is to utilize the water where it falls by means of appropriate land treatments and *in-situ* (*i.e.* at the same place) moisture conservation practices.

Some of these include off season tillage, mulching, dead furrows key line cultivation, compartmental bunding, graded border strips, and interplot water harvesting. Off-season tillage helps the rain water to enter into the soil profile more effectively and in addition, helps in weed control. Such a practice in alfisols (i.e. highly leached red brown soils) of the Telengana region in the state of Andhra Pradesh increased the sorghum yield by 43 per cent. Off-season tillage, however, is not suggested for aridisols (i.e. sandy soils of the desert regions) as this would accelerate wind erosion. Deep ploughing (> 120 cm), once every two or three years, also helps to increase crop yield in soils with a hard subsoil below the plough layer.

Mulching is a well-known practice for minimizing the movement of moisture from the soil into the atmosphere. Poor infiltration of rain water is a problem in vertisols (*i.e.* heavy black soils), which often leads to water stagnation. Vertical mulching (*e.g.* placing sorghum stubbles in trenches 40 cm deep, 15 cm wide, and protruding 10 cm above the ground level) has been found to enhance available soil moisture by 4-5 cm and to increase the grain yield of sorghum by 40-50 percent in vertisols at several locations in India. The making of "idle furrows", also called "dead furrows" at each 3.6 m interval helps in improvising water retention. Such treatment at Anatapur in South India was found to increase the grain yield of groundnut by about 10 per cent.

Ridge and furrow *in-situ* water harvesting is yet another promising system for the inceptisols (*i.e.* soils with poor horizon development as a in alluvial soils) of sub-humid areas. Maize is planted on the ridges and rice in furrows, thus meeting the varying water requirements of crops. The water requirement of rice is met by the water that is harvested into the furrows. Yield increases of about 18-21 per cent for both crops can be obtained by this system, as compared to flat sowing.

The inter row water harvesting system is a practice suitable for light textured soils found in the arid region. Significant yield advantages have been recorded with the pearl millet crop by adopting this practice, in combination with the recycling of the water for supplemental irrigation.

Runoff harvesting and recycling

Rainwater can be utilized by collecting in small dug out ponds excess runoff during high intensity rainfalls and then recycling it for supplemental irrigation at critical stages of crop growth. This water also can be used for pre-sowing irrigation of a *rabi* crop in potential double cropping regions. A *rabi* crop is a crop grown during the post-rainy season (*i.e.* October-March) on stored soil moisture. Extensive studies have been undertaken on these aspects both in arid and semiarid areas. Experiments on catchment size and slope, size of the pond, effective sealants, water lifting devices, and choice of crops have been conducted both in alfisol and vertisol regions.

Existing water harvesting system suffers from the fact that the donor catchment (*i.e.* actual area that is contributing the runoff) does not benefit from the runoff water. In a modified system with small dugout ponds, the water can be used for the donor catchment. Studies have shown that appropriate pond size varies with rainfall, ranging from 200 cu m to 3000 cu m. Seepage losses pose a major problem in these dugouts. Though natural silting minimizes seepage losses in large tanks, it causes a reduction in storage capacity in small dugouts. Among the sealants evaluated, cement plus soil in a 1:8 ratio was found to be reasonably efficient and cost effective. But an ideal sealant has not been found yet that is both efficacious and cost effective.

Among the water application methods tried, an irrigating alternative furrow system was found to be more effective than use of the furrow method and flooding. Despite the clear demonstration of benefits of *in situ* moisture conservation and runoff recycling in both the experimental plots and the operational research projects, these practices have not yet been applied on a wide scale by the farmers. There are many constraints, both financial and technological, but the most critical constraints is the fact that farm holdings in

India are very small and fragmented. The average size of an operating farm is about 1.8 hectares, which limits the application of conservation methods. Therefore, it is now realized that water conservation development projects should be undertaken on a watershed basis. The impact of water resource development can be better seen when improved practices are adopted in a watershed as a whole (*i.e.* a unit of land and a drainage area contributing runoff water to a common collecting point).

Soil and water conservation measures

There are quite a few options in the area of soil and moisture conservation. In most of the situations, contour cultivation and contour sowing will serve as good measures for soil and water conservation. In other situations where slope is more than 1.5 per cent, mechanical measures like contour bunding in low rainfall areas will be appropriate. In deep black soils where water stagnation could be a problem, ridge and furrow, or a broad bed and furrow (BBF) land configurations will be desirable. In soils having a hard pan or compact layer, deep ploughing once in 3 to 4 years may have to be resorted to for ensuring better infiltration of water and root growth.

Incorporation of crop residues, use of surface mulches, vertical mulching, impounding of runoff water in a dug out pond etc. are other soil and moisture conservation measures. In arid areas, setting up of shelter belts / wind breaks is an effective measure of soil conservation.

It is now established that contour farming parallel to graded bunds / key lines helps in conserving by both soil and water. For *in-situ* moisture conservation, furrows are made with a plough after every row, 3-4 weeks after sowing - an operation which is incidental to interculture done in dryland crops. Experience gained in the Operational Research Project at Hyderabad has revealed an additional advantage of about 15 per cent in the yield of sorghum and castor, on account of furrowing as a part of interculture operation done on contour. The practice is easily acceptable by small farmers as it does not involve any additional cost.

As a first step, simple agronomical practices like off-season tillage across the slope, key line cultivation, flat sowing and ridging later, strip cropping, intercropping system have to by adopted. These practices can be easily adopted without much additional cost. Later, mechanical structures may be introduced after gaining full confidence of the farmers. The emphasis should be on in situ moisture conservation measures. Other soil management practices that have been developed for specific situations include compaction of highly permeable soils of low retentively, chiseling soils having layers of high mechanical impedance at shallow depths, raised/sunken bed technology for slowly permeable flat soils for regions receiving more than 1000 mm seasonal rainfall, etc.

In the past, mechanical structures were considered to be the panacea for all soil erosion problems. We never thought of vegetative barriers which could be as efficacious as mechanical structures. Where the slope is steep and the gradient is high, there is no substitute for mechanical structures. There are situations where the soil erosion problems are not that severe and the gradient in the waterway is not that high, vegetative barriers could be established with advantage. At times, we may have to combine, the mechanical drop structures with vegetative barriers. The whole approach will depend on the situation.

Moisture conservation in crop production

Soil moisture conservation management associated with timely operations plays a vital role in dryland areas in water is not limited in quantity only but available within a specified period. Hence there is need to conserve the moisture to keep moisture supply throughout the crop life. Under dryland conditions, soil moisture conservation and management associated with timely operations play a very vital role in crop production since soil moisture is not only limited in quantity but it is also available within specified but uncertain periods.

Methods of moisture conservation in dryland areas

Rainfall water can be conserved in following ways

A) By storing more of rainfall in root zone

1. Tillage
2. Contour farming or bunding
3. Vertical mulching
4. Subsurface barrier
5. Addition of pond sediments of organic matter
6. Addition of gypsum.

B) By checking loss of water through evapotranspiration

1. Mulching
2. Weed control
3. Use of antitranspirants
4. Crop thinning.

C) By water harvesting

1. Water harvesting *in situ*

   i) Inter row water harvesting

   ii) Inter plot water harvesting

2. Water harvesting for recycling

A) Increasing water storage in root zone.

1. Tillage: Tillage opens the soil make it friable and thus it helps in reducing runoff and increases infiltration of rainwater in this way it conserves the maximum amount of rainwater. But it is to be taken into account that whether we should go for deep tillage or shallow. Deep tillage is better than shallow tillage. Because it conserve more moisture and helps in control of weeds.

2. Contour farming bunding: On slopy lands, when tillage operation like ploughing, seed bed preparation, seeding, interculture operations are performed across the slope as nearly on contours as possible. It creates series of miniature barriers to water and thus reduces runoff as well as soil loss.

3. Vertical mulching: Vertical mulching is a recent innovation which is found to be helpful *in-situ* conservation of moisture for increased crop yields. This has been specially found to be suitable to black soils of Deccan plateau whose intake rate is very low. It consist of Jowar stubbles keep in trenches of 40 cm deep, 15 cm wide protruding 10 cm above ground level. Such trenches spaced at 4-5 meters increased crop yields by 400-500% in drought years and 40-50% in normal years ever control.

4. Sub-surface barrier: The practice is suitable for sandy soils where water is lossed through percolation. The placement of subsurface barrier of asphalt (2 mm thick at 60 cm depth) resulted in increased uptake of N by bajra plants. These were associated with higher WUE and yield of crop.

5. Additional of pond sediments of organic matter

6. Application of gypsum

In alkali-soils application of gypsum resulted in increased infiltration rates

B) By checking evapotranspiration losses of water

1. Mulching

With respect to dryland mulch is any material placed at soil surface with a view to conserve the soil water. These may be various types such as - straw or residue mulch, soil mulch, plastic mulch, and chemical mulch.

a) Straw mulching can help in conserving soil moisture in following ways -

   i) By checking loss of water from soil through process of evaporation because it reduces heating of soil from radiation and reducing wind speeds near soils surface.

   ii) By reducing runoff and increasing infiltration of water.

   iii) By checking weed growth.

b) Soil mulching: Surface mulch of dry soil 5-8 cm deep by obstructing the rise of water to the surface through capillary action effectively reduced loss of water as compared with a soil having an undisturbed surface. Following kinds experiment and conclusion, a soil mulch for moisture conservation became standard procedure in dry regions and dicta such as two cultivation can replace one irrigation, became very popular.

c) Chemical mulching : It has been found that Hexadecanol, a long chain alcohol mixed with surface 1/4 inch of the soil reduced evaporation by 43%, this material which is resistant to microbial activity remained effective for more than a year. The surface layer of treated soil dried out more rapidly that of untreated soil, creating a diffusional barrier to evaporation.

2. Weed control

Since time immemorial, controlling weeds has been known to be one of the most effective means of increasing the amount of water available to the crops and therefore, of increasing WUE. Weed frequently transpires greater amount of water per unit of dry matter produced than do the plants with which they grown in association. It has been estimated the amount of water saved by eliminating weeds in a maize field is equivalent to providing an entire irrigation at a time of maximum need.

3. Use of antitranspirants

An antitranspirant is any material applied on transpiring plant surface with the aim of reducing water loss from the plant. Antitranspirants can reduce transpiration in following ways and discussed briefly in chapters 5.

a) By increasing leaf resistance to water vapour loss.

b) By reducing the net energy uptake by the leaves, through increasing leaf reflectance.

c) Reducing crop growth by growth retardents.

84   Rainfed Agriculture

Antitranspirant may be of four types.

i) Stomata closure : *e.g.* PMA (Phenyl murcuric acetate)

ii) Film forming : *e.g.* Plastic and waxy material.

iii) Light reflecting type : *e.g.* Kaolin.

iv) Growth retardant : *e.g.* CCC (2 chloro-ethyl trimethyl ammonium chloride)

4. Crop thinning

Optimum plant population is one of the important practices for getting good yield under limited moisture conditions. Crop thinning may be a mid season correction for dryland agriculture. Thinning of plant population by removing very third row was found to be advantageous to mitigate the moisture stress.

(C) Water harvesting for reuse

Water harvesting is a technology of runoff farming, which aims at increase runoff and decreasing infiltration rate of rain water. It is not only suitable in arid and semi arid regions but also in sub humid regions where annual rainfall is sufficiently high but along dry spells do occur during crop growth.

Water harvesting is of two types

i) Water harvesting *in situ.*

    a) Inter row water harvesting.

    b) Inter plot water harvesting.

ii) Water harvesting for recycling

The same has been discussed earlier in this chapter in brief.

5

Principles of Intercropping; Cropping Systems/Intercropping Systems in Rainfed Agriculture; Mulches and Antitranspirants

Intercropping is growing two or more crops simultaneously on the same piece of land with a definite row pattern. The crops may or may not be sown or harvested at one time.

Advantages of intercropping

Intercropping provides yield advantages as compared to sole cropping. These are not by means of costly input but by the simple expedient of growing crops together.

1. It provides greater surety and stability of higher yield over different seasons.

2. It economizes the space and time of cultivating two or more component crops of comparable agronomic practices grown separately.

3. It helps to restore soil fertility when legumes are included as component crop.

4. It utilizes a greater total volume of both below and above ground environment.

5. It helps to avoid intra-crop competition and thus a higher number of crop plants can be grown per unit area.

6. It helps in better use of growth resources.

7. It also helps in better control of weeds pests and diseases.

86 Rainfed Agriculture

8. It also controls erosion through providing continuous leaf cover over the ground surface.

9. It is the small farmers of limited means who is most likely to benefit.

Disadvantages of intercropping

1. Yields decrease because of adverse competition effect.

2. Allelopathic effect.

3. It may create obstruction in free use of machines for intercultural operations, particularly where the component crops have different requirements for fertilizer, herbicides, pesticides etc.

4. Large farmers with adequate resources may likely to get less benefit out of intercropping.

5. Intercropping management is difficult under high degree of mechanisation.

6. Intercropping may reduce qualitative and generative yields through it may provide a higher quantitative yield.

Criteria for successful intercropping

1. The time of peak nutrients demand of component crops should not overlap.

2. Competition for light should be minimum among the component crops.

3. Complementarity should exists between the component crops.

4. The differences in maturity of component crops should be atleast 30 days.

Intercropping was originally practiced as an insurance against crop failure under rainfed conditions. The main objective of intercropping is higher productivity per unit area in addition to stability in production. Inter cropping system utilizes resources efficiently and their productivity in increased.

Types of intercropping

Mixed intercropping: Growing two or more crops simultaneously with no distinct row arrangement. Also referred to mixed cropping.

Row intercropping: Growing two or more crops simultaneously where one or more crops are planted in rows. Often simply refered to as intercropping.

Strip intercropping: Growing two or more crops simultaneously in strips wide enough to permit independent cultivation but narrow enough for the crops to interact agronomically.

Principles of Intercropping 87

Relay intercropping: Growing two or more crops simultaneously during the part of the life cycle of each. A second crop in planted after the first crop has reached its reproductive stage of growth but before it is ready for harvest often simply referred as relay cropping.

Intercropping under dryland areas

In India large cropped area under dryland conditions (97 m ha) which is 72% of the total cropped area and it contribute near about 42% to the national food basket. So it becomes necessary to grow intercrops within the main crop so that farmer should get good yield even under stress or water scarcity condition De (1989) reported that 20-30 weeks length of effective cropping season in Jhansi, Kovilpatti, Hyderabad, Udaipur, Solapur, Agra, Anand and Akola centres intercropping should be done. Intercropping also practiced where soil moisture storage capacity is 100-150 mm and rainfall is about 625-750 mm.

Where as Subramamniam and Singh (1983) suggested that intercropping system should developed for improving crop production, when the rainfall is between 500 to 700mm with a distinct period of moisture surplus. They also suggested that the future intercropping systems in dry lands of India should be cereal+legumes and the objectives are to:-

1. Minimise fertilizer use

2. Minimise pest and disease incidence in legumes

3. Produce balanced foods

4. Provide legume fodder for cattle

5. Take advantages of extended growing season.

Some of the examples of intercropping at different center of dryland research given by them are presented below:

Important intercropping systems in rainfed areas

Region	Intercropping System
Varanasi	Pearlmillet + Black gram
Agra	Mustard + Chickpea
Jhansi	Sorghum + Pigeonpea
Dehradun	Maize + Soyabean
Rewa	Sorghum + Pigeonpea
Bhubaneswar	Maize + Pigeonpea

88   Rainfed Agriculture

Singh *et al.* (1990) reported that intercropping system sorghum + safflower (2:1) at Bellary, sorghum + greengram and sorghum + chickpea at Kovilpatti gave good returns than sole crop. They also reported that pigeonpea + sunflower at Solapur, soyabean + pigeonpea at Indore, pigeonpea + sesamum at Arjia (Rajasthan), sorghum + pigeonpea at Jhansi, barley + chickpea at Agra, barley + lentil at Varanasi, pearlmillet + greengram at Dantiwada, pigeonpea + cowpea (fodder) at Hisar, grass based intercropping at Jodhpur gave more returns as compared with sole crops.

Region	System	Ratio
Vidarbha region of Maharashtra	Sorghum + Greengram	1:1
	Sorghum + Pigeonpea	2:1
Malwa plateau of MP	Sorghum + Pigeonpea	2:1
Black soil of S-E Rajasthan	Sorghum+ Greengram	1:1
	Sorghum + Pigeonpea	1:1
Deccan region of Maharashtra	Pearl millet + Pigeonpea	2:1
Medium black soil of	Pearl millet + Pigeonpea	4:1
Rajkot region	Groundnut + Pigeonpea	6:1
Sub-humid red soil region	Rice+Pea	4:2
of Orrisa		
Hisar region	Pearl millet + Clusterbean	2:1

Suitable intercrop for different dry farming region

Region	System
Hisar	Pearl millet + Cowpea
	Pearl millet + Greengram
Agra	Pearl millet + Greengram
Bangalore	Pigeonpea + Cowpea
Akola	Sorghum + Greengram
Ranchi	Pigeonpea + Rice

Intercropping and risk distribution

System of intercropping is age old practice and in a way cover the principle of risk distribution agronomy with a particular reference to drought areas with growing season of 20-30 weeks are generally most suited for intercropping system. These areas receive 625-800 mm of rainfall with stored water being 15-20 cm in the root profile.

Choice of inter cropping system should depend on pattern of rainfall and there are three possibility as stated below:

Principles of Intercropping 89

1. Rainfall more uncertain in early part of season

2. Rainfall more uncertain in later part of season

3. Rainfall more or less uniformly distributed.

For areas with uncertain rains in early part of season (i.e. Bangalore region) early planting of the deep rooted drought tolerant crops like Pigeon pea might be useful followed by planting of other component crop. Then the high yielding system of suitable crops need to be compared to the modified intercropping system for areas with uncertain rainfall in the later part of season the companion crop should be shorter in duration than the base crop *e.g.* sorghum+greengram, in areas where rainfall is more or less uniformly distributed. It would be Ideal to take suitable base crop and a companion crop of either longer or shorter duration depending on the growing season. In each of these cases it would always better to develop the system over the best base crop of the region on these principles, different inter cropping system should be tested and evolved.

Based on research conducted for over a decade in different parts of India under the auspices of the All India Co-ordinated Research Project on Dryland Agriculture, the most production intercropping systems have been indentified for different centres and are presented below:

Intercropping systems for different centres in India

Centre		Systems	Ratio
Hyderabad	Sorghum	Sorghum + Redgram	2:1
Akola	Based	Sorghum + Greengram	1:1
Indore		Sorghum + Blackgram	1:1
Udaipur		Sorghum + Redgram	2:1
		Sorghum + Redgram	2:1
		Sorghum + Soyabean	2:2
		Sorghum + Redgram	1:1
Sholapur	Pearlmillet	Pearlmillet + Redgram	2:1
Bijapur	Based	Pearlmillet + Redgram	2:1
Rajkot		Pearlmillet + Redgram	2:1
Udaipur		Maize + Redgram	1:1
Indore	Maize	Maize + Soyabean	1:1
Dehradun	Based	Maize + Soyabean	8:2
Bhubaneswar		Maize + Ktthi	1:1
Ranchi		Maize + Redgram	2:1
Rakh Diliansar		Maize + Greengram	1:1

Bhubaneshwar	Rice	Rice + Redgram	4:2
Ranchi	Based	Rice + Redgram	4:2
Anantapur	Groundnut	Groundnut + Redgram	5:1
Bangalore	Based	Ground nut + Castor	5:1
Rajkot		Groundnut + Redgram	4:1
		Ground nut + Redgram	6:1
		Ground nut + Castor	6:1
Bangalore	Ragi	Ragi + Leucaena	3:1
Varanasi	Based	Ragi + Horsegram	2:1
Bijapur	Redgram	Redgram + Blackgram	1:1
Sholapur	Based	Redgram + Setaria	2:1
		Redgram + Setaria	2:1
Jodhpur	Castor	Castor + Cowpea	1:1
Dantiwada	Based	Castor + Cowpea (F)	1:1
Hyderabad		Castor + Sorghum	1:2
		Castor + Sorghum	1:1
Jodhpur	Fodder	Cenchrus + Cluster bean (F)	2:2
Anand	Based	Dicanthium + Cluster bean (F) Mixture	
		Cenchrus + Cluster bean (Grain)	2:2
Hisar	Pearlmillet	Pearlmillet + Cluster bean	2:1
	Based	Pearl millet + Cowpea	2:1
		Pearl millet + Greengram	2:1

Reducing water losses through mulches and use of antitranspirants

Dryland agriculture is practiced in most of the arid and semi arid areas. The main problem of these regions is low rainfall and meagre availability of irrigation water. A major portion of rainfall is lost in unproductive losses like evaporation, deep drainage, and seepage, there-by leaving a very small amount of water for crop use. The crop-water use can be enhanced and water use efficiency (WUE) increased by controlling unproductive losses of water. Water use efficiency is the yield of marketable crop produced per unit of water used in evapotranspiration. It can be expressed as WUE = Y /ET, where Y = yield of marketable crop and ET = evapotranspiration or seasonal water use. The term evapotranspiration (ET) and water use (WU) are generally considered identical. These are, however, different in the sense that the former term refers to loss of water in both evaporation from the bare soil and transpiration from the crop, while 'the latter is the beneficial utilization of water for producing dry matter and seed. Therefore, for achieving higher water use efficiency the

Principles of Intercropping 91

evapotranspiration losses need to be minimized and the water efficiently utilized for producing higher yields. This is particularly important in dry areas where water is both a scarce and precious commodity. This warrants a discussion on water losses and their control for achieving higher water use efficiency.

Evaporation

Evaporation is a process of losing water from a moist or wet surface to the atmosphere due to vapour pressure differences. From the moist soil surface the water is lost at about the same rate as from a free water surface having the same exposure and temperature. Harrold *et al.* (1959) reported that during the growing period of an annual crop about half the evaporated water comes from the soil. It, however, depends upon the type of soil, climatic conditions, the crop and its stage of growth. In arid regions, the loss of water due to evaporation from the soil is relatively low, the main component of losing plant transpiration. It is because the soil dries quickly in such areas and acts as a mulch.

Evaporation process

Three conditions are necessary for evaporation to take place. First, there must be a continuous supply of heat to meet the latent heat requirement (590 cal/g of water evaporated at 15°C). This can come from the evaporating body or from outside as radiated or advective energy. Second, the vapour pressure in the atmosphere over the evaporating body must remain lower than the vapour pressure gradient between the body and the atmosphere and the vapour pressure must be transported away by diffusion or convection or both. These two conditions are influenced by meteorological factors such as temperature, humidity, wind velocity and radiations *etc*. The third condition relates to the supply of water through the body to the evaporating surface. This condition depends upon the amount, energy and the conducting properties of the evaporating body. Therefore, the evaporation from a soil surface or drying of soil depends upon climatic factors as well as the conducting properties of soil from lower moist layers to the evaporating site. The evaporation or drying of the soil takes place in three stages. The stage one takes place when the soil is wet and is controlled by external meteorological factors and soil surface conditions. Stage second is governed by the rate of supply of water to the evaporating surface. In stage third soil surface becomes extremely dry and the movement of water is in vapour phase. This stage persists for a longer period. The rate of loss of water in this stage is extremely low. In arid areas the first stage persists for a very short time but the second stage prolongs depending upon the soil moisture conditions and its transmission properties.

Evaporation control

Evaporation from the soil surface can be controlled by (i) reducing external evaporativity, (ii) reducing energy supply to the evaporating site, (iii) Decreasing the, conductivity / diffusivity of the, soil, and (iv) reducing the potential or the, force deriving water upwards through the profile. Based on these principles various measures have been devised and are discussed as under.

Shelterbelts/Shelter barriers

In arid and semi arid areas wind is one of the most important factors responsible for increased evaporation and loss of water from soil. Hot dry winds quickly dry the soil surface and create vapour pressure gradient for continuous vaporization. Shelter-belts or shelter barriers are one or multi rows of trees, shrubs and other plants planted in the field or field boundaries across the wind direction for reducing external evaporativity. The importance and utility or shelter belts in controlling wind erosion, reducing evaporative demand and protecting crops from hot desiccating winds particularly in arid and semi arid areas has been widely recognised by many workers in the world. In India, different types of shelterbelts for farmers' fields and roadside plantations in the arid areas of western Rajasthan and also discussed their effectiveness in reducing wind speed and controlling wind erosion. It has been observed that vegetative barriers as shelter-belts placed in tile path of wind reduces its velocity near the ground by exerting a drag on the wind and deflecting the wind stream. The effectiveness of shelterbelts in reducing wind velocity depends on many factors such as wind velocity itself, direction, shape, width, tree height and density etc.

Surface mulching

Mulching is a practice of covering the surface of soil with plastics, organic materials like straw, grass, and stone *etc.* reduce evaporation, to keep down weeds and also to moderate wide fluctuations in diurnal soil temperatures. It controls external evaporativity and also reduces energy supply to the evaporat-ing site by cutting off solar radiation falling on the ground. Its main function is limited to controlling first stage of drying which helps in improved moisture status, reduced soil temperature, besides checking seedling mortality and improving crop stand. Besides the above, mulching helps in increasing downward movement of water. Its storage deep in the profile escapes evaporation due to reduction in thermal gradients and exchange of vapours. The effectiveness of mulches in conserving moisture has generally been found to be higher under more frequency of rainfall, drought conditions and also during early periods of plant growth when canopy cover remains scanty.

Tillage

Tillage is a practice of loosening soil, killing weeds and increasing porosity with the help of some tool or implement. These attributes, in turn, affect soil-water-relationships, aeration status, thermal characteristics and the mechanical impedance to the root penetration. Though tillage increases moisture loss from the soil surface by increasing exposed area and the free exchange of vapours, the dry layer thus created acts as a mulch and restricts the upward movement of water to the evaporating site by reducing diffusivity gradients. Though the water loss from the surface is higher, the cumulative loss from soil profile is reduced with this practice. This is particularly observed in medium sandy loam type of soils. In arid soils the beneficial effects of tillage are specially derived from increased water recharge, reduced crop-weed competition, better root growth and proliferation, improved soil thermal conditions, leading increased water use and water use efficiency.

Transpiration and its control

Transpiration is a process of water loss from the plant surface about 99% of water absorbed by the roots is lost to the atmosphere as transpiration. The loss is mainly from the stomatal pores present on the plant leaves. As the stomata are the focal points on the plant leaf surface where the vapour pressure gradient between the leaf, surface and the atmosphere is the steepest, therefore, these are important sites for the application of antitranspirants. Antitranspirants can be defined as the materials which decrease water loss from the plant leaves by reducing the size or number of stomatal openings, decreasing thereby, the rate of diffusion of water vapours. Two most important points in the use of antitranspirants are: (i) the application of antitranspirants should restrict the water loss from the leaf surface without reducing photosynthesis, as carbondioxide diffuses through stomata and is necessary for photosynthesis, (ii) transpiration causes cooling of the leaf surface and the use of antitranspirants should not com-pletely stop transpiration and thus raise the leaf temperature.

Antitranspirants can be effective in two ways., viz., (i) through films that coat the leaf surface, and (ii) chemicals that close the stomata. Synthetic films that are sprayed on the leaf surface should ideally be resistant to water and completely permeable to carbondioxide. This is, however, very difficult to achieve. Antitranspirants such as phenyl mercuric acetate and certain alkenyl succinic acids act by inhibiting stomatal opening. Film forming antitranspirants (waxy or plastic emulsions) produce an external physical barrier to retard the escape of water vapour from plants. Silicones penetrate the leaf and may act directly on the wet cell walls and white reflecting materials (white wash or kaolinite) lower leaf temperature and reduce the vapour pressure gradient from leaf to atmosphere.

Role of antitranspirants

The main role of antitranspirants is to con-serve water which otherwise is lost through transpiration. Since efficient crop production depends upon timely availability of water, any water conservation practice which enhances the availability should be useful. this is particularly important in arid areas where droughts are frequent and crop growth suffers due to short availability of water. In irrigated farming the interval between two irrigations can be increased by using antitranspirants and thus water can be saved. .

In many cases the primary purpose of using antitranspirants is not water conservation per se but improvement of plant performance by increasing plant water potential. Antitranspirants can be used to control temporary shocks to plants during transplanting. The use of stomata inhibiting or film forming antitranspirants for sorghum grown with limited irrigation with 5-17% increase in yield. Although antitranspirants have been found to reduce both transpiration and photosynthesis, they have been found quite useful in conserving water for making it available during stress periods for sustained growth and production specially in arid and semi arid areas.

Deep percolation and its control

Sandy soils occupy an, important place in many arid and semi arid areas. These soils are single grained, have poor moisture retention and storage capacity, allow heavy leaching and per colation losses and are thus infertile and unproductive. Soils of arid areas of western Rajasthan have been found to contain 58 to 71 % fine sand and 24-28% coarse sand. These soils retain about 6 to 9% moisture at 0.1 bar tension which comes to about 90 to 140 mm of water per metre of soil profile. During periods of high rainfall or high rainfall intensity about 40-50% rainfall is lost as deep percolation retaining thereby, low amounts of water in the root zone for plant use. Though the deep percolation water in the strict sense is not a loss as this goes to recharge the ground water and is available to perennial vegetation, yet it is not available to the field crops or plants with shallow root system. Therefore, research efforts were made to intercept this water for utilization of the crops/orchard plants.

Sub-surface moisture barrier

Subsurface moisture barrier is an artificial layer of a material with less permeability placed at a suitable depth in profile for intercepting the free flow of water. The water thus retained over the surface of barrier is low tension water and is easily available to the plants during the periods of drought. The barrier concept originated from the observation that sandy soils with clay sub soil are world's most productive soils. Though sub surface barriers were found quite useful, their large scale use could not be achieved because of non availability of suitable machinery for placement, large scale soil disturbance and high cost involved. However, sub surface barrier of bentonite clay and pond sediments were effectively used for orchard crops and establishment of trees in the arid areas, an integrated use of circular water harvesting technique with sub surface moisture barrier of pond sediments was found highly effective for the growth and yield of *Zizyphus* (Ber) and establishment of *Acacia tortilis* (Babul) and *Prosopis cineraria* (Khejri) plants and reducing their mortality under rainfed arid conditions of western Rajasthan.

Soil compaction

Compaction is a process of bringing soil particles closer to attain higher microporosity and thereby increased water retention in soil. Though higher levels of compaction have been reported to produce high mechanical impedance to root penetration and thus adversely affect plant growth, low levels create favourable conditions by way of increased moisture and nutrient retention.

Agroforestry system

Agroforestry is yet another technique to control deep percolation. Different trees like *Prosopis cineraria, Acacia senegal, Acacia nilotica, Zizyphus rotundifolia, Zizyphus mauritiana* if planted in agricultural fields of the arid areas, utilize the deep percolation water efficiently, because of. their deep root system. The system thus helps in meeting grain, fuelwood and fodder requirements beside protecting and preserving the environment.

Role of antitranspirant in dryland agriculture

Globally 2/3 of earth surface is occupied by water but paradoxically only 1 % ground water is vital for subsistence of life on earth. Crops grown exclusively on rainwater are classified on dry land crops. Out of total land on earth 1/4 of the surface (11 %) is cultivable. 28% is rainfed area, 22 % without soil microbes, 10% is waterlogged, 23% of soil is without nutrient and 6 % land is covered with ice. In Africa, only 6% total land is under dryland. In India, particularly is the country where cover of 70% land under rainfed farming and it is very essential to manage every drop of water received through rains. "Antitranspirant is defined as any material applied to transpiring plant surface to reducing water loss from the plant". The possibility of reducing plant transpiration by chemical without material reducing photosynthesis is of great practical importance in arid region. Where crop production is limited by lack of water, however the potential uses of antitranspirants are not restricted to water conservation but may also includes the maintains of more favourable water balances in plants. Particularly in situation at plant growth stages. The practical use of antitranspirants in field crops involves decreasing water loss from leaves by reducing the size/ number of stomatal opening and thus decreasing the rate of diffusion of moisture vapour. However, photosynthesis in the leaf depends on supply of CO_2 diffusing into the stomatal cavity. The loss of water through plant needs to be controlled because only a small portion of water taken up by plant roots incorporated into the plant body. Thus the very low efficiency in water utilization by green plants often a tremendous challenge for research in methods of decreasing water loss from vegetations. Another methods that of CO_2 enrichment are of the air is intended tom directly increase the rate of photosynthesis. Generally when plant photosynthesizing affectively and stomata's are widely open. The mesophyll resistance may be considerable greater than the stomatal resistance. The total impact of such an increase in respiration on the evaporation rate will be greater than the photosynthesis rate. Antitranspirants when sprayed or applied through soil can benefit the crop under adverse rainfall. Drought reduces the crop yield by 0-100% depending upon severity. Normally it is reduced by 20-60 %. Crop productivity is dependent upon how fast a plant can recover after a stress of 6-10 days. Prolonged drought can be drastically reducing yield to zero level. Drought during the critical phenological phase like flowering and grain development is highly detrimental. Transpiration is said to be unavoidable evil but it has several function to attend in crop cycle. For one tone of food, about 1000 tones of water is needed. Cereal and legumes need 400-500 liters of water/ kg of grain, while fruit and vegetables require 1000 liters of water/ kg of food produced.

According to Waggoner, the transpiration is represented by:

$$T = \frac{\blacktriangle H_2O}{r_s + r_a}$$

where, T – Transpiration

and $\blacktriangle H_2O$ – Difference in vapour concentration between stomatal cavity free air outside

r_s – Stomatal resistance

r_a – Resistance of boundary air layer

Thus the transpiration can be decreased by :

Principles of Intercropping 97

1. Decreasing $\blacktriangle$ H_2O in the above equation
2. Increasing stomatal resistance
3. Increasing boundary air layer resistance.

This increase/decrease can be done by use of antitranspirants

Water requirement of some crops and critical stages of growth

Crop	Water required (mm)	Critical period (Days after sowing)
Sorghum	400-500	28-30
Bajra	250-300	50-55
Maize	400-420	30-45
Wheat	400-450	60-65
Groundnut	400-450	40-45
Sunflower	300-350	30-35
Safflower	250-300	50-60
Gram	250-300	60-65
Cotton	550-600	80-90
Pigeonpea	500-550	90-100

Types of antitranspirants

1. Stomata closing type: Most of the transpiration occurs through the stomata on the leaf surface. Some fungicide like Phenyl mercuric acetate (PMA) and herbicides like atrazine in low concentration serve as antitranspirants by inducing the stomatal closing. These might reduce the photosynthesis also simultaneously. PMA was found to decrease transpiration to greater degree then the photosynthesis in a number of plants the results of field studied have not be particularly increasing. Glyceryl half ester of decenyl succinic acid (GLOSA). Paraquat, Atrazine, cycocyl are the examples. Stomatal opening is regulated by various sensors like, water, CO_2, light and hormones. The opening is strongly controlled by hydroactive mechanism while other sensors are hydropassive and is mediated through RWC of guard cell chloroplast. The antitranspirants curtail the water requirement but it lead to proportionate reduction in photosynthesis.

1. The chemical should be impermeable to water and permeable to CO_2
2. It should be non – toxic to the plants and animals
3. It should have longer retention and greater specificity
4. It should be degrade with solar UV radiation
5. It should be relatively inexpensive and easily available.

98   Rainfed Agriculture

Recently, methanol (10%) is shown to have antitranspirant property and reduces stomatal conductance and transpiration.

PMA and ABA are capable of reducing stomatal aperture. The exact mechanism is not known but it is thought that they may alter the permeability of the guard cells. Thus they decrease transpiration. The problems with these are:

1. By reduction in transpiration leaf temperature becomes high and this will tend to increase transpiration and respiration and thus may decrease WUE.

2. Decrease in photosynthesis rate.

3. These chemicals are toxic and we are interested in non-toxic material which have long lasting effects.

2) Film forming type: These antitranspirants when sprayed on plants provides a physical barrier on the stomata. These will escape the loss of water vapour from the leaves. But the problem with these is they hinder the exchange of CO_2 & O_2 which ultimately affect the photosynthesis rate.

3) Reflectant type: These are white material, which form a coating on the leaves and increase the leaf reflectance (albedo). By reflecting the radiation, they reduce the leaf temperature and vapor pressure gradient from leaf to atmosphere and thus reduce transpiration. Application of 5% kaolin spray reduces transpiration losses. Kaoline and Bordeaux mixture (white ash, $CuSO_4$) are important chemicals under reflecting type. The objective is to apply a reflective pigment which increase the albedo and thus decrease the net radiation load and leaf temperature.

4) Growth retardants: These chemicals reduce shoot growth and increase root growth and thus enable the plants to resist drought. They may also induce stomatal closer. Cycocel is one such chemical useful for improving water status of the plant.

Effect of antitranspirants (ATS) in crop

Adequate available water during the period from tillering through grain filling in cereals is necessary for production of high yields in semi arid regions. ATS may be used to retard the rate of soil water depletion.

Physiological effects

Physiological effect are mostly centralised on photosynthesis and transpiration (TR) because of its on gas exchange. The ATS reduce the rate of TR by limiting the vapor exchange through the stomata. It has also found that adverse effect on photosynthesis.

Effects of yield and yield attributes

Out of two antitranspirants Kaoline and PMA, the Kaoline increase number of panicle, 1000 grain of weight, spikes length, number of grains/ spikes in wheat more than PMA. However, increase in yield attributes of raya due to application of both PMA and Kaoline. Researchers reported significant increase in yield of spring wheat grown in green house due to PMA.

List of various types of antitranspirants

S.No.	Name of chemical	Types/ class	Mode of application
1.	Adol-52 (cetyl alcohol)*	Soil water binding	Soil
2.	OED (Octadecanol) *	-do-	Soil
3.	S-800*	plastic	Spray
4	Alkanyl suyccinic acid	stomata	Spray
S	Sodium-Hydroxydecanol sulphonate	-do-	Spray
6	Monoethyl etherof decenyl succnic	-do-	Spray
7.	acid	Stomata closing	Spray
8.	8-Hydoxyquinoline sulphonate	Reflecting	Spray
9.	Kaoline	Stomata closing	Spray
10.	Abscissic acid (ABA)	-do-	Spray
11.	Cycocel	-do-	Spray
12.	Ascorbic acid	-do-	Spray
13.	Phenyl mercuric acetate (PMA)	Reflecting	Spray
14.	Calcium bicarbonate	-do-	Spray
15.	China clay	Stomata closing	Spray
16.	TIBA	Plastic	Spray
17.	Waxol (wax)	Plastic	Spray
18.	Mobileaf (Trade name)	Stomata closing	Spray
19.	Folicot (Trade name)	-do-	Spray
20.	Hico 110 R (Trade name)	Plastic	Spray
21.	Polyethyline vinyl acryline	Stomata closing	Spray
22.	Phosphon	Stomata closing	Spray
23.	Paclobutrazol	Stomata closing	Spray
24.	Methanol	Stomata closing	Spray
25.	34-D	Stomata closing	Spray

Water use efficiency

As expected, water use efficiency (WUE) is increased by use of antitranspirants, especially under moisture stress conditions. PMA increased WUE of wheat at 1×10^{-4} M but decreased it at higher concentrations (7.5 x 10^{-4} M). Giri *et al.* (1983) also noted increased relative leaf water content and WUE of wheat after using PMA or Kaolin. Antitranspirant spray on potato plants during tuber enlargement reduces plant water use and increases the yield of large tubers. Increased relative leaf water content and increased WUE of chillies sprayed with PMA or Kaolin at low soil moisture contents have been recorded. Effects of filmforming and silicone antitranspirants were also found suitable for herbaceous plant species.

Moisture conservation

Evapotranspiration of wheat plants sprayed with 1×10^{-4} M PMA at 40, 60, and 80 per cent soil moisture depletion was 44.92. 42.52 and 38.35 cm, respectively. Kaolin-sprayed wheat plants grown under dryland saved 36 cm water. Kaolin-sprayed barleys maintained more available moisture in the soil. The consumptive value and the soil moisture depletion pattern were markedly influenced. Kaolin used in conjugation with PMA reduced daily evapotranspiration significantly. The reduction lasted for two weeks under clear weather. Irrigations, one at crown root initiation and the second at (lowering, produced similar yields to irrigation at all four stages. Spraying at crown root initiation and flowering with irrigation at jointing and grain filling, gave significantly higher yields than irrigation at crown root initiation and flowering with spraying at jointing and grain filling.

Variability in response to antitranspirants

All crop species do not respond similarly to antitranspirant treatments, even the varieties differ in their response. Season, region, stage of crop growth, and management practices also influence the effect of antitranspirants, both qualitatively and quantitatively. The efficacy of the antitranspirant (hydrasyl) varied widely with plant species and depended on both physiological and morphological factors. Triadimetron reduced transpiration and increased yield of soybeans and peas but not of wheat. Transpiration was reduced five to six days after treatment, the effect being greater in oats than in wheat. Maize and barley plants were sprayed with 2 per cent emulsions of paraffin oil or textol at three- to four-leaf stage. Paraffin oil gave lower water losses in maize while textol gave lower losses in barley.

Effects on growth and yield

Although the rate of photosynthesis is slightly reduced after the use of antitranspirants the plants' water economy is improved and wilting is avoided. Plants continue to grow at a lower rate than well irrigated plants, but at a higher rate than unsprayed plants. Thus, the growth and yield of antitranspirant-sprayed plants are improved under rainfed dryland conditions. Kaolin spray on wheat increased grain yield due to increased tiller survival, longer ears and greater spikelet fertility. The number of shrivelled grains per ear was decreased. A combination of mulching and transpiration suppression brought about a saving of one irrigation on clay loan soil and at least two irrigations on a loamy sand soil.

Role of mulches in dryland agriculture

Mulch is any material at the soil surface that was grown and maintained in place; any material grown, but modified before placement and any material processed, manufactured or transported before placement. The "materials include crop residues, leaves, dippings, bark, manure, paper plastic films, petroleum products, gravel and coal. The practice of applying mulches to soil is possibly as old as agriculture itself. Ancient Romans placed stones on the soil surface to conserve water and Chinese used pebbles from streambeds. The current trend under mechanized agriculture is to use crop residues as mulches on cultivated areas, usually in conservation tillage systems. An alternative is to use transported and manufactured materials as mulches for some high value crops. Mulches are used for various reasons but water conservation and erosion control are undoubtedly the most important for agriculture in dryland (semi-arid and arid) regions. While the effectiveness of mulches for water conservation is highly variable, mulches when properly managed definitely wind and water erosion can be controlled. Other reasons for mulching include soil temperature moderation, soil nutrient effects, soil salinity control, soil structure improvement, crop quality control and weed control. Considerable research on the use of mulches has been conducted in these aspects in dryland areas.

Effects of mulches on soil properties and conditions

Mulches affect numerous soil properties and conditions either directly or indirectly. Among these affected are soil water content through runoff control, increased infiltration, decreased evaporation, and weed control; soil temperature through radiation shielding, heat conduction and trapping and evaporative cooling, soil nutrients through organic matter additions, differential nitrification, and mineral solubility, soil structure, soil biological regime through organic matter additions, microbial and soil fauna populations, and plant root distributions, soil erodibility and soil salinity through leaching and evaporation control. Probably the greatest importance, however, for agriculture in dry regions are soil water, temperature, structure and salinity. The effects of mulches, on these properties and conditions are briefed below:

Soil water

By their very nature, semiarid and arid regions generally receive inadequate precipitation for good crop production. Furthermore, much of the precipitation that is received is lost by runoff and/or evaporation. Also, some water that enters the soil may evaporate before a crop is planted. Although, precipitation is limited and erratic and the evaporation potential is high, good yields can be obtained in semiarid and arid regions if most of the precipitation that occurs is effectively conserved and subsequently used for crop production. Numerous studies have been conducted to determine the influence of mulches on soil water storage, content, and evaporation. Many different materials have been used, but besides crop residues, one of the most widely researched one is plastic film (included are polyethylene, polyvinyl chloride and similar films).

Although, plastic films generally resulted in higher water content and reduced evaporation as compared with bare soil, they are relatively expensive and difficult to manage, especially under large scale field conditions for low value crops. Consequently, less expensive and more readily applicable materials have been sought. Petroleum and asphalt sprays and resins have received considerable attention. These materials affected soil water in a manner similar to plastic films. Their effectiveness depended on the amount of surface covered. Most research with mulches has involved crop residues and other plant waste products (*e.g.*, straw, stover, leaves, corn cobs, sawdust, and woodchips). These materials are cheap and often readily available, and they permit water to enter the soil readily. When maintained at adequate levels, these materials increase infiltration and soil water contents and reduce evaporation.

Soil temperature

When surface coverage was adequate, mulches generally resulted in greater soil water contents and lower evaporation than bare soil; however, the effects of mulches on soil temperatures were highly variable. Colour of plastic mulches greatly affected soil temperature. White or reflective plastic decreased temperatures. Adams (1962) stated that clear plastic result in higher temperatures because the soil directly absorbs most of the energy from incoming solar radiation. The effectiveness of clear plastic for increasing soil temperatures increased with aging of the plastic. With black plastic, the soil received only a portion of the incoming energy absorbed by the film. Apparently, degree of contact between soil and plastic influences the extent of temperature increase under a black plastic. White and reflective plastics resulted in cooler temperatures because they did not absorb the incoming radiation.

Petroleum spray and resin mulches consistently resulted in higher temperatures than bare soil. The maximum difference usually occurred during the hottest part of the day. At night, the mulches had slight or no effect on soil temperatures. These materials (petroleum, asphalt, manure) absorbed the incoming solar radiation and, because of their close contact with the soil, readily conducted heat to the soil. Hence, the consistently higher temperatures with these mulches as compared with the variable influence of black plastic.

Mulches of plant materials (*e.g.*, straw, stover, leaves) reduced soil temperatures. The combined effects of radiation in transpiration and evaporative cooling, were responsible for the lower temperatures under this type of mulch. The 0-9 t/ha range reduced soil temperature at a 10 cm depth by 0.18 °C.

A group of miscellaneous mulching materials variously, and somewhat unpredictably, influenced soil temperatures. Black granular materials (*e.g.*, coke, bitumen) increased temperatures, whereas lighter coloured gravel reduced temperatures as compared with bare soil. The dark materials absorbed more radiation than the lighter coloured gravel; hence, the higher temperatures with the black materials.

Soil structure

In dry regions, precipitation frequently occurs in high intensity storms. When falling raindrops strike bare soil, soil particles are dispersed and surface sealing may occur, thus, reducing infiltration. Consequently, water that could be conserved for plant use is lost by runoff. The dispersed soil at the surface often forms a hard crust when dry, which may adversely affect seed germination. Seedling emergence, plant growth and soil aeration. Beneficial effects of surface mulches on soil structure result primarily from mulches absorbing the energy of falling raindrops, thus reducing soil dispersion and surface sealing. Infiltration rates generally are maintained, and subsequent crusting is reduced. Other benefits possibly result from greater microbe and fauna activity in soil, greater root proliferation, and the cushioning effect, which reduces compaction due to tractor, implement, or animal traffic.

The degree of protection provided at the soil surface is related to surface coverage provided by the mulch. Plastic films prevent raindrops from striking the soil and, thus prevent dispersion. However, unless special provisions are made, they also prevent water infiltration. Less compaction was also ascribed to a plastic mulch.

Petroleum, bitumen, coke, and desert pavement mulches reduced surface crusting and soil density. The materials, except possibly petroleum, undoubtedly also resulted in greater water infiltration. Although effective for reducing crusting, these materials may not provide the necessary binding substances for improved soil aggregation. However, the surface remains adequately covered, stable surface aggregates are less important than where the soil is partly or completely exposed.

Crop residues and other similar mulches, when present or applied at adequate levels, have maintained high infiltration rates. The high infiltration rates resulted from less surface sealing due to the protection against falling raindrops afforded by the mulches. Decomposition products of the mulches resulted in improved soil aggregation with respect to aggregate size, number and stability. Although, aggregation was improved, the aggregates were not sufficiently stable to withstand the impact of water drops without a protective cover.

Conservation tillage systems that maintain most crop residues on the surface have given variable resulted with respect to soil structure. Many authors reported larger water stable aggregates and lower bulk densities as a result of stubble mulch tillage when compared to clean tillage.

Soil salinity

Many soils in semi-arid and arid regions have a high salt content. Because some of the salts are readily soluble in water, they move with the water. The salts can be removed from the soil by leaching if precipitation is adequate. However, due to limited precipitation, they often move only to a limited depth and readily return to the surface with capillary flow as the soil water evaporates. Under such conditions, susceptible plants may be severely injured by the salts, especially at the germination and seedling stages. With irrigation, the salt problem may be aggravated by applying water having a high salt content and by applying insufficient water to move the salts well below the plant root zone. The management of saline soil has been widely investigated.

Because salts readily move with soil water, and practice that maintains infiltration rates and reduces subsequent evaporation should moderate the adverse effects due to soil salinity. Damage due to salts is most severe at germination and plant seedling stages; hence, any practice that reduces the salt content in the seed zone should be beneficial for plant establishment. While complete soil covers may be too expensive and impractical for extensive areas, strip mulches (e.g. plastic films, petroleum sprays) that maintain a higher water content in and reduce evaporation from the seed zone should reduce the salinity hazard. Additional benefits should result if the mulches continue to reduce evaporation and the crop yield and quality are improved due to salinity control.

A soil salinity problem associated with increased water conservation, as with stubble mulch tillage under fallow conditions, is the saline-seep problem in some areas. The problem develops when water refills the soil profile and deeper layers at upslope positions. As the water flows down slope through the soil, it carries with dissolved salts. The saline water seeps to the surface at downslope positions, adversely affecting crop production at the seep sites. The most effective solution to the saline-seep problem is to use as much of the current precipitation as possible for crop production before the water percolates beyond the root zone. Forage crops, especially perennials that have deep root systems, are effective for reducing the problem.

Effects of mulches on erosion control

Because of low residue production and dry soil surface conditions semi-arid and arid regions are highly susceptible to wind erosion. Water erosion also can be a problem because the precipitation frequently occurs as intense storms and the surface is inadequately protected by vegetation to effectively retard runoff. Therefore, mulches are important in semi-arid and arid regions for reducing erosion. The ease by which soil particles are moved by wind and water is related to particle size and to wind or water velocity. Although particles greater than 0.84 m in diameter are generally considered non-erodible by wind, water may cause erosion of particles of almost any size. Decomposition products of some mulching materials may increase soil aggregation, thus reducing the susceptibility of soil to erosion. The chief value of mulches with regard to erosion control, however, is not their effect on soil structure but their moderating the wind and water forces at the soil surface. Additional benefits for water erosion control result from increased water infiltration rates and reduced surface runoff due to mulches.

A wind erosion equation has been developed which relates soil loss by wind to the

(1) Soil erodibility index measured in terms of soil clods greater than 0.84 mm in diameter;

(2) Climatic factor measured in terms of wind velocity and surface water content;

(3) Soil surface roughness;

(4) Unsheltered field width; and

(5) Vegetative cover.

The stubble-mulch farming system, widely used in North America, was developed primarily for wind erosion control. The value of stubble- mulch tillage for erosion control is widely recognized. In general, erosion by wind decreased as the amounts of crop residues maintained on the surface were increased by stubble mulch tillage. Wind erosion was even more effectively controlled where higher amounts of residue were maintained on the surface by chemical fallow or no tillage practices.

Standing residues were more effective for wind erosion control than flattened residues; but to shield against the erosive action of moving water, the mulch must reduce water flow at the soil water interface. Hence, a maximum amount of the material should be in contact with the soil surface. In general, erosion was reduced as the amount of mulch was increased. However, if the mulch was not in adequate contact with the soil, erosion could occur under the mulch. In recent years, no tillage systems have received considerable attention. With these systems, crop residues are maintained on the surface, weeds are controlled with chemicals, and subsequent crops are seeded with no more soil disturbance than necessary to place the seeds in the soil. Excellent control of erosion by wind and water has been obtained.

Effects of mulches on plants

The effects of mulches on plants are operative through their effects on soil water, temperature, structure, salinity, and erosion. Critical periods in the life cycle of a plant are germination, emergence and seedling establishment. For germination to occur, viable seeds must be placed in a favourable environment with respect to water supply, temperature, and aeration. After germination, the seedlings must emerge and become established. Due to their small size and tenderness, seedlings can easily be adversely affected by an unfavourable environment. Mulches can aid germination, emergence, and seedling growth and yield moderating or improving the soil and aerial environment to which the seeds and seedlings are subjected. Higher soil water contents and reduced evaporation were major reasons for improved germination, emergence, and seedling growth due to mulches. Straw, petroleum, gravel, stones, and plastic films increased germination and early growth; but the plastic films had to be slit or removed for continued growth. Higher soil temperatures due to mulches also improved germination, emergence and seedling growth. The higher temperatures, either alone or in combination with higher water contents, were also effective in promoting later plant growth and hastening plant maturity. Early seeding of sweet corn, made possible by using a clear plastic mulch, resulted in the corn being ready for harvest two weeks before corn seeded in bare soil.

Surface mulch, which maintained a more constant soil temperature than other materials after irrigation were effective for improving growth of early potato *(Solanum tuberosum* L.). In other cases, lower temperatures due to mulches reduced germination and early plant growth. However later growth sometimes was better on mulched than on bare soil due to improved water conditions under the mulch. Many of the reports have indicated higher yields when crops were grown with mulches, rather than without mulches, with most mulching materials being effective for increasing yields. These ranged from about 50 to 300 per cent.

Yield response due to mulches, although primarily related to soil water and temperature, was affected also by better plant populations, reduced root rot. Undoubtedly, mulch effects on soil structure, nutrients, microbial activity, root distributions, and other factors also played an important role in the higher yields. Reduced crop yields due to mulches have been related to some specific conditions, generally soil temperature. Corn apparently is very sensitive to low temperatures under straw mulch early in the season when temperatures normally limit plant growth. The adverse effects of low temperatures were reflected in lower yields in more northern regions, but were largely overcome in the more southern regions.

6

Concept of Watershed Resource Management, Problems, Approaches and Components

Watershed management

Watershed is defined as a geohydrological unit draining to a common point by a system of drains. All lands on earth are part of one watershed or other. Watershed is thus the land and water area, which contributes runoff to a common point.

A watershed is an area of land and water bounded by a drainage divide within which the surface runoff collects and flows out of the watershed through a single outlet into a lager river or lake.

Types of watershed

Watershed is classified depending upon the size, drainage, shape and land use pattern:

1) Macro watershed (> 50,000 ha)

2) Sub-watershed (10,000 to 50,000 ha)

3) Milli-watershed (1000 to10000 ha)

4) Micro watershed (100 to 1000 ha)

5) Mini watershed (1-100 ha)

(a) This is the process of guiding and organising land and other resource usage in a watershed.

(b) Ensuring the sustenance of the environment mainly the soil and water

110   Rainfed Agriculture

resources need to recognise the interrelationships between, land use, soil-water, and slope of terrain.

(c) Unifying focus in watershed management is in how various human activities affect the relationship between water and other natural resources. Provides a basis for actions concerning the development and conservation.

Objectives of watershed management

The different objectives of watershed management programmes are:

1. To control damaging runoff and degradation and thereby conservation of soil and water.

2. To manage and utilize the runoff water for useful purpose.

3. To protect, conserve and improve the land of watershed for more efficient and sustained production.

4. To protect and enhance the water resource originating in the watershed.

5. To check soil erosion and to reduce the effect of sediment yield on the watershed.

6. To rehabilitate the deteriorating lands.

7. To moderate the floods peaks at downstream areas.

8. To increase infiltration of rainwater.

9. To improve and increase the production of timbers, fodder and wild life resource.

10. To enhance the ground water recharge, wherever applicable.

Factors affecting watershed management

a) Watershed characters

i)   Size and shape

ii) Topography

iii) Soils

iv) Relief

b) Climatic characteristic

i.   Precipitation

ii. Amount and intensity of rainfall

c) Watershed operation

d) Land use pattern

i. Vegetative cover

ii. Density

e) Social status of inhabitants

f) Water resource and their capabilities

Watershed management practices

1. In terms of purpose

i. To increase infiltration

ii. To increase water holding capacity

iii. To prevent soil erosion

2. Method and accomplishment; in brief various control measures are:

a. Vegetative measures (Agronomical measures)

i. Strip cropping

ii. Pasture cropping

iii. Grass land farming

iv. Range land measures

b. Engineering measures (Structural practices)

i. Contour bunding

ii. Terracing

iii. Construction of earthern embankment

iv. Construction of check dams

v. Construction of farm ponds

vi. Construction of diversion

vii. Gully controlling structure

viii. Providing vegetative and stone barriers

ix. Construction of silt tanks

Rainwater harvesting is the main component of watershed management. Some of the watershed management structures are as follows.

Broad beds and furrows

a. Function

To control erosion and to conserve soil moisture in the soil during rainy days

b. General information

The broad bed and furrow system is laid within the field boundaries. The land levels taken and it is laid using either animal drawn or tractor drawn ridgers

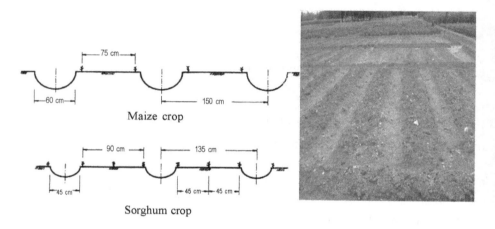

Broad beds and furrows

c. Cost

Approximate cost for laying beds and furrows is Rs.1800 / ha

d. Salient features

i. Conserves soil moisture in dryland

ii. Controls soil erosion

iii. Acts as a drainage channel during heavy rainy days.

2. Contour bund

a. Function

To intercept the runoff flowing down the slope by an embankment

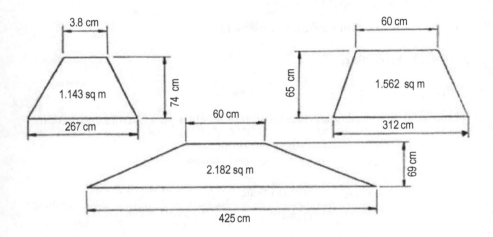

Contour Bunding

b. General information

It helps to control runoff velocity. The embankment may be closed or open, surplus arrangements are provided wherever necessary.

c. Cost

Approximate cost of laying contour bund is Rs.1400 / ha.

d. Salient features

  i. It can be adopted on all soils
 ii. It can be laid upto 6% slopes
iii. It helps to retain moisture in the field.

3. Bench terracing

a. Function

It helps to bring sloping land into different level strips to enable cultivation

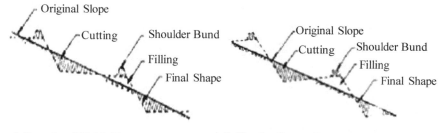

1. Level and Table Top   2. Sloping Inward

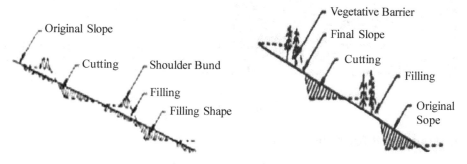

3. Sloping Outward

4 Puertorico type terrace

b. General information

It consists of construction of step like fields along contours by half cutting and half filling. Original slope is converted into level fields. The vertical & horizontal intervals are decided based on level slope

c. Cost

Approximate cost for laying the terrace is Rs.5000 / ha

d. Salient features

i. Suitable for hilly regions

ii. The benches may be inward sloping to drain off excess water

iii. The outward sloping benches will help to reduce the existing steep slope to mild one

iv. It is adopted in soils with slopes greater than 6%.

4. Microcatchments for sloping lands

a. Function

It is useful for in-situ moisture conservation and erosion control for tree crops

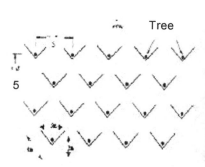 

Micro-catchments

b. General information

Technique	Storage capacity per unit (m³)	Annual runoff contribution to soil moisture additional water stored per ha (m³)	Estimated surface Runoff control (%)
Triangular catchments (V-Bunds)	13	5200	80
Crescent bunds	10.2	4080	73

c. Cost

Technique	Cost/ha (Rs.)
Triangular catchments (V-Bunds)	6000-7000
Crescent bounds	2500-3000

d. Salient features

1. Slope ranges from 2 –8%
2. Soil type – light to moderate texture
3. *In-situ* moisture conservation with staggered planting

4. Suitable for dry land horticulture and agroforestry
5. Bund height – 30 to 45 cm

5. Check dam

Salient features

1. A low weir normally constructed across the gullies
2. Constructed on small streams and long gullies formed by erosive activity of flood water
3. It cuts the velocity and reduces erosive activity
4. The stored water improves soil moisture of the adjoining area and allows percolation to recharge the aquifers
5. Spacing between the check dams water spread of one should be beyond the water spread of the other
6. Height depends on the bank height, varies from a metre to 3 metre and length varies from less than 3m to 10m
7. Cost varies from Rs. 40000/- to Rs. 100000/- per unit.

6. Percolation Pond

Function

To augment the ground water recharge

Salient features

1. Shallow depression created at lower portions in a natural or diverted stream course.

2. Preferable under gentle sloping stream where narrow valley exists.

3. Located in soils of permeable nature.

4. Adaptable where 20-30 ground water wells for irrigation exist within the zone of influence about 800 – 900m.

5. Minimum capacity may be around 5000 m^3 for the sake of economy.

6. Also act as silt detention reservoir.

7. Cost varies from Rs. 60000 to 150000 per unit.

Principles of watershed management

The main principles of watershed management based on resource conservation, resource generation and resource utilization, are:

1. Utilizing the land according to its capability.

2. Protecting productive top soil.

3. Reducing siltation hazards in storage tanks and reservoirs and lower fertile lands.

4. Maintaining adequate vegetation cover on soil surface throughout the year.

5. *In-situ* conservation of rain water.

6. Safe diversion of excess water to storage points through vegetative waterways.

7. Stabilisation of gullies by providing checks at specified intervals and thereby increasing ground water recharge.

8. Increasing cropping intensity and land equivalent ratio through intercropping and sequence cropping.

9. Safe utilisation of marginal lands through alternate land use systems with agriculture horticulture forestry pasture systems with varied options and combinations.

10. Water harvesting for supplemental and offseason irrigation.

11. Maximizing agricultural productivity at convenient locations per unit area per unit time and per unit of water.

Concept of Watershed Resource Management, Problems   119

12. Ensuring sustainability of the ecosystem befitting the man-animal-plant water complex.

13. Maximizing the combined income from the inter related and dynamic crop livestock tree labour complex over years.

14. Stabilising total income and to cut down risks during aberrant weather situations.

15. Improving infrastructural facilities with regard to storage, transportation and marketing of the agricultural produce.

16. Setting up of small scale agro industries; and.

17. Improving the socio-economic status of the farmers.

Components of watershed management programme

The main components of watershed management are:

1. Water resource improvement

2. Soil water conservation and water harvesting

3. Crop management

4. Alternate land use systems and integration of livestock in farming system

5. Land treatment in non arable lands.

Successful implementation of the above components in the watersheds would lead to the development of livestock, poultry, pisciculture and other associated and allied activities, finally resulting in improved standard of living of the farmers and rural masses.

Water resource improvement

Various water resources should be improved in the watershed command areas. The water from these resources should be collected in the watershed which can be used for the crop at critical growth stages.

Soil water conservation and water harvesting

Soil and water conservation measures coupled with water harvesting help to improve the moisture availability in the soil profile and surface water availability for supplemental and off-season irrigation. The interventions through soil conservation measures have greater role to play in transferring a part of surface

120   Rainfed Agriculture

water to ground water by recharge. Based on the nature and type of hydraulic barriers and their cost, the conservation measures in arable lands can be divided into following three categories.

1. Hardware treatments
2. Medium software treatments
3. Software treatments.

Hardware treatments

Hardware treatments are generally of permanent type, provided for improvement of relief, physiography and drainage features of watershed. These are erected with major Government support with the purpose to check soil erosion, regulate overland flow and reduce peak flow. At times, they are imposed to completely divert the upland runoff from running into downstream fertile lands. Different hardware treatment are explained below.

Waterways: Quite often, high runoff volumes are observed, even in drylands, particularly in watersheds located in sloppy areas. Such runoff water should, therefore, be chanelised through a few waterways; as far as possible some of the existing waterways may be developed. Waterways draining larger areas may be designed on hydrologic and hydraulic considerations. No doubt, at places, sizeable land strip may be occupied by such waterways, but his may be found highly advantageous for the survival of many multi-purpose useful trees when planted along the waterways. Depending upon land situations alignment of small size waterway may be adjusted with field boundaries; at times, such arrangements are specially useful for safe guarding crops against over stagnation of rain water in black soil areas. Such water course may also be safeguarded against erosion by providing mechanical checks and also by raising vegetation which may be useful fodder for animals. In some situations, waterways may also have a small section side bund with necessary openings at water entry points.

Bunds: These are low height earthen embankments constructed across the slope in cultivated lands after deciding location of waterways. The bunds function to intercept runoff, increase infiltration opportunity time and dispose excess rainfall safely. Such bunds can be either contour bunds or graded bunds; many a time, these bunds are adjusted with field boundaries, if deviation from grade or contour is not too much; no doubt spacing depends on many engineering considerations.

Contour bunds are recommended in dry farming areas with light textured soils of slopes upto 6 per cent and where annual rainfall does not exceed 600 mm. They are designed for an expected runoff of 24 hours duration and 10 years frequency. Surplussing arrangements (waste weirs) are provided to dispose of the excess runoff beyond the design storage. The cross section area of contour bunds follows the depth and type of soil; however, $0.5m^2$ is minimum.

Graded bunds are constructed in medium to high rainfall areas in poor permeable soils (vertisol), having 2 to 6 per cent slope. They are also quite suitable for the soils having crust formation tendency like red *'chalka'* soils of Telengana region of Andhra Pradesh. By and large, graded bunds with 0.3 to 0.5 in section are constructed with longitudinal gradient of 0.2 to 0.4 per cent depending on the site condition (Singh et al., 1990). Graded bunds particularly are made with the following objectives:

1. To reduce runoff

2. To reduce soil loss

3. To divert runoff.

Thus, graded bunds along with waterways and water harvesting structures not only check soil erosion, but they also provide an ideal rain water management system for many watershed situations.

For making bunds, required soil may generally be taken from lower side. Sometimes, excavations and pits for making bunds in deep black soil areas particularly are made on upper side. But these may be got leveled or ploughed up as early as possible after bund constriction. Invariably barring a few situations borrow pits should be made on lower side at the places where the bunds drain into waterways. In real field situations, bunds are covered with natural grass and in most cases, these are dependable sources of forage. In addition, quite often some trees also come up in their close vicinity, to the liking of cultivators. In some places loose stones are easily available; hence stone bunds are quite advisable in such pockets.

Terracing: Cultivated lands having slopes of 10 per cent particularly in hilly areas should be put under bench terracing by converting the lands into series of platforms. The width of bench terrace depends on the land slope and the permissible depth of cut. Bench terracing is very much effective in reducing soil erosion in hilly areas. At Ootacamund, it has helped in bringing down the annual erosion rate from 39 t/ha to less than 1.0 t/ha on 25 per cent sloping lands. At places where scattered stones are available, loose stone walls can be made to act as risers for bench terrace construction.

122 Rainfed Agriculture

Zing terrace consists of a contributing (donor) area with mild slope and a receiving levelled area at lower side half to one third of the donor area. They are particularly suited for areas having medium to deep soils.

Medium software treatments

Medium softwares are also provided particularity as inter bund treatments, where field sizes are large and conventional bunds are constructed along field boundaries. Such treatments are usually of semi-permanent type and are adopted to minimise the velocity of overland flow. They may need major initiative from cultivators in addition to some grants from the Government side. Such measures may last 2 to 10 years; vegetation component and land configurations at times may also provide some direct returns on short term basis. However, they need to be modified at times to maintain effectiveness for erosion control and moisture conservation, and also to minimise risk of providing shelter to harmful pests in and around these measures and in certain situations like taking up corrective measures to avoid too much spread of introduced vegetation and being counter productive.

Small section/key line bunds: A small section bund may be created across the slope at half of the vertical bund spacing. Such bunds can be nearly 0.1 m^2 in section; may be renovated at an interval of 2 to 3 years.

Strip leveling: About 4 to 5 in strip of land above the bund across the major land slope may be levelled for the purpose. Similarly, one or more strips can be created at midlength of slope. Such strips can be created by running blade harrows after ploughing the field with mould board or disc plough. Such minor rough levelling programme may be taken up after every 2 to 4 years.

Live beds: One or two live beds (2-3 m wide) can be created either on contour or on grade in the inter bund space. The vegetation on the beds may be according to the liking of cultivator, it can be annual or perennial or a combination of both.

Vegetative/live barriers: One or two barriers of close growing grasses or legumes can be created alongbunds as well as at the mid length of slope to filter the run off water or slow down, overland flow. *Khus* could be one of such vegetations. Several other promising grass species that serve as valuable fodder for cattle are being explored as an alternative to *khus* grass. A miniature bund at lower side of barrier is recommended to help in the development of live barriers, particularly in the initial stages.

Software treatments

A mention was made that hardware type land treatments are useful for safe runoff disposal and similarly, medium software are essential to slow down the velocity of overland flow in cultivated fields. However, on several occasions, these are found inadequate in attaining equitable moisture distribution for crop growth. In such cases software treatments are taken up for ensuring fairly uniform soil moisture for satisfactory crop performances. By and large, software treatments are temporary in nature; in that case these are required to be remade or renovated every year. The entire cost of applying such treatments is to be met by the farmers. Because of favourable economics, a few of these treatments have gained wide acceptance in the recent years.

Contour farming: Contour farming is one of the easiest and most effective and low cost method of controlling erosion and conserving moisture. With contour farming, tillage operations are carried out along contours. It creates numerous ridges and furrows for harvesting sizeable amount of runoff inside the field itself.

Compartmental bunds: Compartmental bunds, converting the area into square/rectangular parallels are useful for temporary impounding of water for improving moisture status of the soil. These are made using bund formers. In medium deep black soil, they are found advantageous in storing the rainfall received during the rainy season in the soil profile there by augmenting the soil moisture for use by rabi crops. The size of the compartments may be fixed' considering the size, of the inter bunded land.

Broad bed and furrows: This technique is especially suited to black soils, where crops are sown on pre formed beds. This system is made before the season and is maintained year after year. The planting is done on the bed. Generally, the depth of each furrow is kept 0.15 m and the inter furrow spacing is maintained at 1.5 m.

Dead furrows: Dead furrows are laid across the land slope in rolling lands, to intercept the runoff. The spacing between dead furrows varies between 2 to 5 m for 4 to 7 crop rows. This system works well in alfisols.

Tillage: Tillage operations help in rain water harvesting and increases infiltration. Off season tillage, in particular, has been found quite useful in most rainfed areas. The tillage operations make the soil receptive to rainfall. This practice is very useful in light soils often prone to crusting.

Mulching: Surface mulching protects soil against beating action to rain drops and it also increases water infiltration into the soil. Further it helps in minimising water evaporation from soil surface. Sometimes dry soil mulch created simply by stirring the soil has been found effective for good performance of crops.

Soil and water conservation in non-arable lands

Control grazing is one of the simplest approaches in reducing soil loss from denuded sloping lands. As seen from the experiences of Operational Research Projects, soil and moisture conservation measures are required even in the non agricultural lands. These practices include contour and staggered trenches and contour furrows. They help in accelerating vegetation establishment, encouraging natural regeneration of species and combating soil erosion. Contour and staggered trenches are usually spaced 10 to 20 m apart across the slope for raising forest species. In general, 0.45 m wide trenches with 0.45 m depth are made at regular intervals. Sometimes in place of trenches, contour furrows are formed at a spacing of 2 to 10 m (Singh *et al.*, 1990). In rocky areas, crescent bunds of loose boulders are constructed in place of trenches. All the above stated measures are useful in slowly regaining the lost forest cover and fertility of most, degraded grasslands; community efforts in the right direction are likely to make major dent in re greening such areas.

Water harvesting measures

Since time immemorial, several kinds of water harvesting structures have been in practice in our country. Recently, local experiences and precedence's were utilised for designing these structures. In order to bring cost effectiveness, efforts are also necessary to harmonize the designing of water harvesting structures based upon traditional wisdom and scientific contemplations. In-fact, farm ponds, surface water tanks and percolation tanks need to be viewed as 'conservation structures in addition to sources for supplemental irrigation. In real watershed situations, ponds maybe located in conjunction with waterways some maybe of a few hundred cubic metre size, whereas couple of structures may be of bigger size. In selected pockets, integrated catchments cum command approach should be followed. To ensure proper storage in light soils, some of the ponds can be lined with sealant materials. All large ponds have to be constructed at Government cost. All such structures are likely to recharge ground water reservoir. When located in proper co cultural hydrologic matrix, in conjunction with other factors, they also help in reducing load on ground water aquifer. It maybe advisable to work out design details of all the required structures in the first phase of the project, no matter construction of a few of these may have to be deferred may be planned to be taken up in drought years to provide employment to the rural folk. The objectives of water harvesting structures are:

1. To store water for supplemental and off season irrigation
2. To act as silt detention structure
3. To recharge the ground water
4. To raise aquaculture species
5. Recreation and allied agricultural uses.

Thus in most situations, in addition to reviving traditional water harvesting systems, other measures are also needed. Some of such typical water harvesting structures adopted in watersheds is as follows:

Minor irrigation tanks: Minor irrigation tanks are constructed across the major streams with canal system for irrigation purpose by constructing low earthen dams. A narrow gorge should be preferred for making the dam in order to keep the ratio of earthwork to storage as minimum. The height of the dam may vary from 5 to 15 m. The tanks are provided with well-designed regular and emergency spill ways for safety against side cutting. In micro watersheds, water harvesting boundaries, similar to small scale irrigation tanks are also recommended. By and large, the water harvesting boundaries are not integrated with extensive canal system.

Farm ponds: Farm ponds are typical water harvesting structures constructed by raising an embankment across the flow direction or by excavating a pit or a combination of both. Dug out ponds in light soil require lining of the sides as well as the bottom with suitable sealants. This is not the case with heavy black soil. Considering the benefit of the harvested water, different type of linings, *viz.*, brick, concrete, HDPE, *etc.* may be used.

Nala bunds and percolation tanks: Nala bunds and percolation tanks are located in the nalas having permeable formations with the primary objective of recharging the ground water. A strict regulation on the silt load entering the downstream reservoirs is an additional advantage of percolation tanks. The percolation tanks encourage the digging up of wells downstream of recharged area for irrigation purposes. The percolation tanks are provided with emergency spillways for safe disposal of flow during floods.

Stop dams: Stopdams are permanent engineering structures constructed for raising the water level in the nala for the purpose of providing life saving irrigation during drought periods. These are located over flat nalas at narrow gorges carrying high discharge of long durations. They are created over stable foundation conditions where hard rocks are encountered. For the stop dam, a site with larger water storage capacity would be desirable.

Crop management

The rainfed areas of our country contribute about 45 per cent of the total food grain production. In the watersheds there is scope of increasing the overall productivity by adopting the integrated farming systems approach. For achieving sustainability in crop production in the watersheds due importance is given to : (i) evolve and follow up simple and low cost crop production technology including improved varieties, (ii) provide alternate crop production technologies to match weather aberrations; and (iii) optimize the use of natural resources like land and water.

Considering the availability of rain water over space and time, different crops have been identified as given in Table below for different rainfall zones.

Agro-ecological conditions in different rainfall zones of India

Mean annual rainfall (mm)	Predominant soil types	Mean annual temperature (^{0}C)	Length of growing season under rainfed condition (days)	Predominant rainfed crops
<400	Sandy	20-30	30-80	Grasses, Pearl millet
		10-20	30-90	Short duration pulses
400-1000	Sandy soils (Aridisols); Red soils (Alfisols and relates soils);	20-30	80-200	Pearlmillet, sorghum, maize, oilseeds, pulses
	Black soils (vertisols and vertic inceptisols)	-	-	Cotton
1000-1800	Alluvial soils (Entisols); Laterite soils (Alfisols and relates soils); Balck soils (vertic and vertic inceptisols)	20-30	200-300	Maize, paddy, wheat, barley, mustard and soybean
> 1800	Sub-montane soils (inceptisols)	15-20	> 300	Paddy, plantation crops

Depending on the water availability periods, examples of some selected crops for rainy season and post rainy season are given in Table below:

Major efficient food crop-based double cropping systems for different rain dependent regions of India (based on water availability periods)

Water availability period (days)	Rainy season crops	Post rainy season crops
110-150	Cowpea/blackgram	Safflower
	Soybean	Mustard/safflower
	Greengram	Chickpea/barley
150-175	Greengram	Sorghum
175-200	Cowpea	Sorghum
	Greengram	Safflower
	Balckgram	Barley/mustard
	Pearlmillet	Chickpea
	Maize	Wheat
		Chickpea/mustard
	Rice	Chickpea
	Sesamum	Chickpea
200-250	Sorghum/groundnut/maize	Safflower
	Soybean	Wheat
		Safflower/chickpea
		Soybean/maize
> 250	Rice/maize/finger millet/groundnut/pearl millet	Wheat/chickpea/linseed/lentil/ horsegram/barley
	Soybean	Wheat
	Pearlmillet	Wheat

Many drylands of the country are not only thirsty, these are hungry as well. Optimum doses of fertilizers are, therefore, recommended for the crops depending upon physico chemical properties of soil. Farmers are advised for proper crop residue management stubbles of the plants being left in the field for decomposition and building up the fertility. However, major breakthrough has yet to be achieved in cropping systems in the watersheds by changing over to organic farming with tree based cropping systems. The research work is being in-itiated on these lines through alternate land use systems.

Alternate land use systems and integration of livestock in farming system

In the watersheds, most of the uplands are degraded and these are also marginal lands with low productivity level. Due to the population pressure, more and more marginal and sub marginal lands are being brought under cultivation. Apart from being uneconomical in the long run (with more inputs), cultivation of such lands, can lead to serious imbalaces in the eco-systems. For such lands, some alternate efficient land use system other than arable cropping would be desirable. In alternate land use system, by encouraging tree and grass components, the demand for food and fodder can be solved to some extent. Thus, stability in production will be achieved along with the safety of environment. As seen from some model watersheds alternate land use systems not only help in generating much needed off season employment but they also help in utilizing off season rain which may otherwise go waste.

In general, land capability classes are the determinant factors for deciding different alternate land use options. By and large, in land classes I to Ill, arable farming can be practiced and beyond that lands should be utilized for forestry and grassland management. For economic considerations, agro forestry options viz. integrating farming with other land uses like horticulture, agriculture and pasture may find acceptance from the farmers. The evaluation of all these systems is under research investigations. The tree component of the system should be selected according to the soil and climatic conditions. The trees should preferably have the characteristics like fast growing with high palatability, good coppicing ability, and drought resistance also having properties like adding organic manure in soil profile. The pasture component in such systems should have the characteristics like growing well even under shade, drought resistance, easy propagation, high palatability, conserving soil and moisture, withstanding over grazing etc. Different legumes like style species and grasses including *Cenchrus* species have been found performing well in many situations.

Dryland horticulture is promising enterprise in slightly favorable soil conditions especially where there is provision of water harvesting. In this hardy fruit species such as ber, pomegranate, custard apple, guava and aonla may have to be preferred over others.

Integrated farming system having the integration of crops with livestock, poultry, fish, mushroom, bee keeping *etc.* should be practiced keeping the resources in view. Different models of farming systems suited to the farming conditions of the individual should be adopted.

7

Drought and Its Management

Low rainfall or failure of monsoon rains is a recurring feature in India. This has been responsible for drought and famines. The word drought, generally, denotes scarcity of water in a region. Though, aridity and drought are due to insufficient water, aridity is a permanent climatic feature and is the culmination of a number of long term process. However, drought is a temporary condition that occurs for a short period due to deficient precipitation for vegetation, river flow, water supply and human consumption. Drought is due to anomaly in atmospheric circulation.

Stresses

External conditions that adversely affect growth, development, or productivity trigger a wide range of plant responses are:

- Altered gene expression
- Cellular metabolism
- Changes in growth rates and
- Crop yields

Types of stress

- Biotic stress - imposed by other organisms
- Abiotic stress - arising from an excess or deficit in the physical or chemical environment. Biotic and abiotic stresses can reduce average plant productivity by 65 to 87%, depending on the crop.

Resistance or sensitivity of plants to stress depends on

- The species
- The genotype
- Development age.

Stress resistance mechanisms

- Avoidance mechanisms prevents exposure to stress
- Tolerance mechanisms permit the plant to withstand stress. Acclimatization alter their physiology in response to stress.

Stresses involving water deficit

Water related stresses could affect plants if the environment contains insufficient water to meet basic needs.

Environmental conditions that can lead to water deficit in the following ways

- Water logging
- Drought
- Hyper saline conditions
- Low or high temperatures
- Transient loss of turgor at mid day.

Factors that can affect the response of a plant to water deficit

- Duration of water deficiency
- The rate of onset if the plant was acclimatized to water stress.

Tolerance to drought and salinity

Osmotic adjustments a biochemical mechanism that help plants acclimatize to dry and saline conditions. Many drought tolerant plants can regulate their solute potentials to compensate for transient or extended periods of water stress by making osmotic adjustments, which results in a net increase in the number of solutes particles present in the plant cell.

Osmotic adjustment : Osmotic adjustment occurs when the concentration of solutes within a plant cell increases to maintain a positive turgor pressure within the cell. The cell actively accumulates solutes and as a result the solute potential drops, promoting the flow of water into cell.

Heat stress

The typical response to heat stress is a decrease in the synthesis of normal proteins, accompanied by an accelerated transcription and translation of new proteins known as heat shock proteins (HSPs).

Drought stress

Control of drought stress in plant is not only very complex, but is also highly influenced by other environmental factors and by the developmental stage of the plant. The effect of drought and heat stress on plant physiology, growth and development is discussed detail in Chapter 13 under "Stress Physiology".

Drought

There is no universally accepted definition of drought. Early workers defined drought as prolonged period without rainfall. Drought is a situation when the actual seasonal rainfall is deficient by more than twice the mean deviation. American Meteorological Society defined drought as a period of abnormally dry weather, sufficiently prolonged for lack of water to cause a severe hydrological imbalance in the area affected.

Prolonged deficiency of soil moisture adversely affected crop growth indicating incidence of agriculture drought. It is the result of imbalance between soil moisture and evapotranspiration needs of an area over a fairly long period as to cause damage to standing crops and to reduce the yields.

Classification of drought

Drought can be classified based on duration and nature of users. In both the classifications, demarcation between the two is not well defined and many a time overlapping of the cause and effect of one on the other is seen. Droughts are classified into four kinds.

Permanent drought

This is characteristic of the desert climate where sparse vegetation growing is adapted to drought and agriculture is possible only be irrigation during entire crop season.

Seasonal drought

This is found in climates with well defined rainy and dry seasons. Most of the arid and semiarid zones fall in this category. Duration of the crop varieties and planting dates should be such that the growing season should fall within rainy season.

Contingent drought

This involves an abnormal failure of rainfall. It may occur almost any where especially in most parts of humid or subhumid climates. It is usually brief, irregular and generally affects only a small area.

Invisible drought

This can occur even when there is frequent rain in an area. When rainfall is inadequate to meet the evapotranspiration losses, the result is borderline water deficiency in soil resulting in less than optimum yield. This occurs usually in humid regions.

Droughts are also classified based on their relevance to the users

Meteorological drought

It is defined as a condition, where the annual precipitation is less than the normal over the area for prolonged period (month, season or year)

Atmospheric drought

It is due to low air humidity, frequently accompanied by hot dry winds. It may occur even under conditions of adequate available soil moisture. Plants growing under favourable soil moisture regime are usually susceptible to atmospheric drought

Hydrological drought

Meteorological drought, when prolonged, results in hydrological drought with depletion of surface water and consequent drying of reservoirs, tanks etc. This is based on water balance and how it affects irrigation as a whole for bringing crops to maturity

Soil drought

This occurs when soil moisture supply lags behind evapotranspiration. It is usually gradual and progressive. Plants can therefore adjust atleast partially, to the increased moisture stress.

Agricultural drought

It is the result of soil moisture stress due to imbalance between available soil moisture and evapotranspiration of a crop. It is usually gradual and progressive. Plants can therefore, adjust at least partly, to the increased soil moisture stress. This situation arises as a consequence of scant precipitation or its uneven distribution both in space and time. It is also usually referred as soil drought.

Relevant definition of agricultural drought appears to be a period of dryness during the crop season, sufficiently prolonged to adversely affect the yield. The extent of yield loss depends on the crop growth stage and the degree of stress. It does not begin when the rain ceases, but actually commences only when the plant roots are not able to obtain the soil moisture rapidly enough to replace evapotranspiration losses. Important causes for agricultural drought are :

- Inadequate precipitation
- Erratic distribution
- Long dry spells in the monsoon
- Late onset of monsoon and
- Early withdrawal of monsoon.

Plant adaptations to growth in dry regions

Levitt (1972) distinguishes between two basic ways in which plants can grow and survive in dry habitat: (a) escaping drought, and (b) actual drought resistance

Escaping drought

Evading the period of drought is the simplest means of adaptation of plants to dry conditions. Many desert plants, the so-called ephemerals germinate at the beginning of the rainy season and have an extremely short growing period, which is confined to the rainy period. Between germination and seed maturity, as few as five to six weeks may suffice. These plants have no mechanism for overcoming moisture stress are therefore not drought resistant.

The presence of germination inhibitors in such seeds serves as an internal raingauge, which permits germination only after an amount of precipitation has fallen that will be sufficient to remove the inhibitor (Went, 1965).

Early maturity

In cultivated crops, the ability of a cultivar to achieve maturity before the soil dries out is the main adaptation to growth in a dry region. However, only very few crops have such a short growing season, that they can be considered similar to the 'ephemerals' in escaping drought. One example is of certain varieties of millets that can produce mature seeds within 60 days from germination.

Drought resistance

Plants that cannot escape periods of drought can adapt to these conditions essentially in two ways:

i. Avoiding stress

Stress avoidance is the ability to maintain a favourable water balance and turgidity even when exposed to drought conditions, thereby avoiding stress and its consequences. Stress avoidance is mainly due to morphological-anatomical characteristics, which themselves are the consequence of the physiological process induced by drought (Levitt, 1972). These xeromorphic characteristics are quantitative and may vary according to the environment conditions

A favourable water balance under drought conditions can be achieved either by : (i) conserving water, by restricting transpiration before or as soon as stress is experienced (the so-called water savers); or, (ii) accelerating water uptake sufficiently, so as to replenish the lost water (water spenders).

The mechanisms for conserving water:-

(1) Stomatal mechanisms: Stomata of different species vary widely in their normal behaviour and range from staying open continuously to remaining closed continuously. Many cereals open their stomata only during a short time in the early morning and remain closed during the rest of the day. There are however differences in this respect between varieties of the same crop, as shown by the following examples : in two varieties of oats, the one more resistant to drought opened its stomata more rapidly in the early morning when moisture stress is at its minimum and photosynthesis can proceed with the least loss of water (Stocker, 1960). An interesting observation made in the course of trials on the effects of soil moisture regime on different wheat varieties was that the stomata of a semi-dwarf variety remained open throughout the day, while those of the tall variety were open for only a few morning hours remaining closed thereafter even under favourable conditions of soil moisture and transpiration (Shimshi & Ephrat, 1970).

However, mechanisms of conserving water based on the closure of stomata will inevitably lead to reduced photosynthesis and may lead to "drought induced starvation injury" (Levitt, 1972).

Drought and Its Management 135

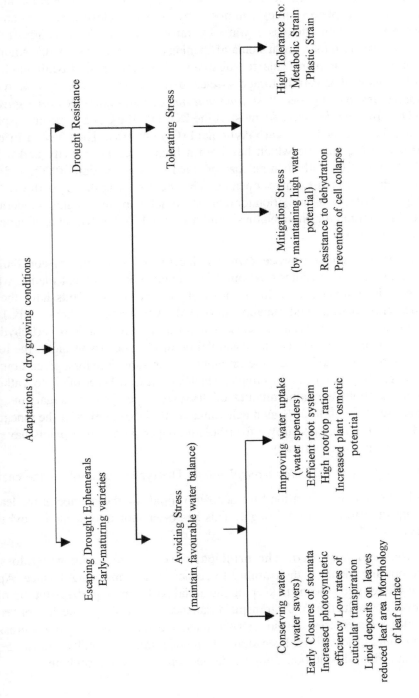

136   Rainfed Agriculture

(2) *Increased photosynthetic efficiency* : One possibility for overcoming the limitations on photosynthesis imposed by stomatal closure as a means for increasing resistance to loss of water by transpiration would be a higher rate of CO_2 assimilation for a given stomatal opening (Hatch & Slack, 1970). A number of important crop plants (maize, sugarcane, sorghum, proso, foxtail and finger millets) as well as certain forage species, bermuda grass (*Gynodon dactylon*), sudan grass, bahia grass (*Paspalum notatum*), rhodes grass, (*Chloris gayana*) and certain *Atriplex* spp., fix most of the CO_2 into the C_4 of malic and aspartic acids; the so-called C_4-dicarboxylic acid (C_4) pathway. They have a primary carboxylating enzyme which has both a high potential activity and a high affective for CO_2 with such an enzyme, high rates of fixation of CO_2 and with relatively restricted stomatal opening. The species using the C_4 pathway have a high rate of CO_2 assimilation for a given stomatal opening, higher temperature and light optima for photosynthesis and higher light saturated rates of apparent photosynthesis.

Another photosynthetic process which reduces water loss without a concomitant reduction in photosynthesis is found in a number of succulent plants, which keep their stomata closed during most of the day. These plants have the so-called crassulacean acid metabolism (CAM) which enables them to fix large amounts of CO_2 as organic acids at night and convert it into carbohydrate during the day. As the primary assimilation of CO_2 occurs at night, the lower temperatures prevailing limit the amount of water transpired for a given stomatal opening. The pineapple, for example, which has the CAM type of photosynthesis, produces about the same amount of dry matter per year as sugarcane, but often uses only 10 to 20% as much water. With the exception of the pineapple and certain *Agave* spp. grown for fibre and paper pulp, CAM plants have not been exploited agriculturally.

(3) *Low rates of cuticular transpiration*: The typical example is the cactus.

(4) Decreasing transpiration by a deposit of lipids on the surface of the leaves on exposure to moderate drought. This has been shown to occur in soybeans (Levitt, 1972).

(5) *Reduced leaf area*: The principal means of reducing water loss of xeromorphic plants is their ability to reduce their transpiring surface. Apart from the common means to keeping the aerial parts small, perhaps the simplest form of this reduction of the transpiring surface is the rolling or curling of leaves at times of water stress, a characteristic phenomenon exhibited by many grasses. The rolling of leaves has been shown to reduce transpiration by almost 55 per cent in semiarid conditions, any by 75 per cent in desert xerophytes.

Drought and Its Management 137

(6) *Leaf surface* : Various morphological characteristics of leaves help to reduce the transpiration rate and may affect survival of plants under drought conditions, leaves with thick cuticle, a waxy surface and with the presence of spines, etc., are common and effective. In certain species, drought stimulates the production of epidermal hairs. The mere presence of pubescence on the leaves is no longer considered as an advantage in reducing transpiration, unless it be thought increasing their albedo.

(7) *Stomatal frequency and location*: A smaller number of stomata can retard the development of water deficits. In certain species, the stomata are located in depression or cavities in the leave, which is a feature that can further reduce transpiration by limiting the impingement of air currents.

(8) *Effect of awns* : Awned varieties of wheat predominate in the drier and warmer regions, and have been found to yield better than awnless ones, especially under drought conditions, though there are exceptions. Awns have chloroplasts and stomata and so can photosynthesize; it has been found that the contribution of the awns to the total dry weight of the kernels was 12% of that by the entire plant.

Accelerating water uptake

(1) *Efficient root systems*: The root systems of drought resistant plants are characterized by a wide variety of apparent adaptations. These are responses to such predominant soil conditions as the duration of soil dryness and the depth that is normally wetter, the presence and nature of soil constituents and the degree of soil salinity. Plants become adapted to dry condition; mainly by developing an extensive root system, rather than by structural modifications of the roots. The concept "extensive root system" also includes, in addition to growth in depth and lateral growth, the density of roots per unit volume of soil and the number of secondary hair roots.

There are considerable differences between cultivated plants in the extent, depth, and efficient, of their root system. Sorghum has nearly twice as many small branch roots per unit length of main roots has maize and it is assumed that this is one of the main reasons for its greater drought resistance. Another adaptation of some xerophytes is their speedy response to a rewetting of the soil by the rapid formation of secondary rootlets ('rain roots') which may appear only two to three hours after a fall of rain (Evenari, 1962). An important characteristic of certain cultivated plants is their ability to respond quickly to fluctuations in moisture content by having roots that grow rapidly towards sources of available moisture. It has been shown experimentally that drought resistance varieties of spring wheat is related to the rapid growth of the primary roots.

Plants are also known to differ with respect to the ability of their root systems to penetrate relatively dry soils. Oats, wheat and barley show little penetration of the soil, at or below wilting point, while two range grasses – side-oats grama (*Bauteloua curtipendula*) and love-grass (*Eragrostis eragrostis*) penetrated the soil quite extensively (Salim *et al.*, 1965).

(2) *High root-to-top ration* : A high root-to-top ration is a very effective means of adaptation of plants to dry conditions, as under such conditions the growth rate of the roots considerably exceeds that of the shoots. The transpiring surface is thereby reduced, while the root system of the individual plant obtains its water from a large volume of soil. An increased root : top ration may actually result in a greater amount of total dry weight of plants grown under dry conditions as compared to similar ones grown with full moisture.

(3) *Differences in osmotic potential of the plants* : Levitt (1958) has calculated that a difference of 0.5% in the soil moisture content that induces permanent wilting, could supply a plant with enough water to keep it alive for 6 days, this could mean, in certain cases the difference between survival and death.

(4) *Conversion of water spenders to water savers* : Because of their increased water absorption, water spenders are characterized by very high rates of transpiration. However, as soon as the absorption rate becomes insufficient to keep us with water loss, the water spenders generally develop some of the characteristics of the water savers.

ii Drought tolerance

When the plant is actually submitted to low water potentials it can show drought tolerance by either mitigating the actual stress induced by the moisture deficiencies, or by showing a high degree of tolerance to these stresses.

Mitigating stress : Adaptations to drought based on mitigating the effects of stress permit the plants to maintain a high internal water potential in spite of drought conditions. They therefore are able to maintain cell turgor and growth, avoid secondary drought induced stress, as well as direct or indirect metabolic injury due to dehydration.

(1) *Resistance to dehydration* : The simplest method of avoiding drought induced damage is by resisting dehydration – preferably to the extent of maintaining turgor, and at least by avoiding cell collapse after loss of turgor.

(2) *Prevention of leaf collapse* : In many instances the principal effect of leaf adaptation to dry conditions, such as thick cell walls, is to prevent wilting or collapse of the leaf, rather than to ensure reduced water loss. This explains why xeromorphic types of leaves frequently transpire more, when amply supplied

with water, than mesomorphic types and thus make possible an efficient use of water during the short periods of wet conditions which are typical for dry environments. As soon as the soil moisture level again becomes critical, the stomata close and the morphological adaptations to reduced water supply mentioned above become operative.

Tolerating stress

(1) *Resistance to metabolic strain* : The greater the elastic dehydration strain. Cell dehydration is elastic and completely reversible up to a point, beyond which it is plastic, irreversible and therefore injurious, the greater the danger of a resulting plastic metabolic strain. When dehydration occurs, there are mechanisms for tolerating an accompanying plastic metabolic strain.

Starvation : The greater the dehydration a plant can tolerate and still keep its stomata open the lower will be its dehydration compensation point. This is the water potential at which rates of photosynthetic CO_2 absorption and respiratory CO_2 evolution are equal, so that net photosynthesis equals zero. This kind of drought tolerance will be due to the specific ability of the guard cells to remain turgid even when the leaf cells are wilted, and is possible due to an increase in the solutes of the guard cells.

Protein loss: The avoidance of protein loss under drought conditions is the net result of two processes : a decreased rate of protein breakdown and or increased rate of synthesis. Young leaves are more resistant to drought than older leaves and this is ascribed to their greater protein. In tests with varieties of winter wheat, for example, water stress caused a greater decreased in RNA content in a nonresistant variety than in the resistant variety.

(2) *Avoidance or tolerance of direct drought-induced plastic strain* : Some higher plants may show little restriction of transpiration during drought and can become air dry without suffering permanent injury. Grama grass, for example, may lose 98.3% of its free water without injury (Oppenheimer, 1960). Exposure to a sublethal stress may cause an increase in resistance so that the plant is able to survive an otherwise lethal stress. This has led to the developing of methods for increasing hardiness of crop plants.

Effects of drought

- Delay or prevention of crop establishment.
- Weakening or destruction of established crop.
- Predisposition of crops to insects and diseases.

140   Rainfed Agriculture

- Alteration of physiological and biochemical metabolism in plants.
- Alteration of quality of grain, forage, fibre, oil etc.
- Drought is often a factor in yield reduction.

Drought management strategies

In our country, irrigated area constitutes 33% and 67% is dryland and rainfed, out of 143.8 m ha of cultivated land. Total cropped area is 180.36 m ha. Dryland agriculture contributes about 45% of national food grain production. After full exploitation of dryland, it may contribute up to 75% of total food grain production. Pulses and oilseeds are mainly grown in such areas. Important commercial crops viz. cotton, castor, groundnut and all coarse grains viz. jowar, bajra, maize crops are rainfed. Problems of dryland agriculture are inadequate, uncertain, uneven rainfall. Drought is major problem, so we go for contingency planning.

Nature of drought

Haryana receives almost 1200-1300 mm of rainfall every year but the rains are erratic in many areas. The probability of rainfall failures and coefficient of variations is quite high in the last weeks of June-July and in the last weeks of September-October. Hence, drought in the state primarily occurs at the start or end of the *kharif* season. In July, upland crops grow to maturity and seedlings for transplanted rice are established. If there is deficient rain, the upland crop, mainly paddy and maize, which provides food security in August-September, is affected. Seedlings for the transplanted paddy start to wilt or become over-mature.

As most of the land is monocropped, the lowland paddy is crucial for employment and food security. A delay in rainfall affects the transplantation of paddy, the major crop in the area. October rains are required for paddy and provide the necessary residual moisture for the *Rabi* crop. The entire state is not generally affected by drought. However, there are areas which are affected frequently. Over a period of time, new areas become affected by drought, and we see that drought has been officially declared in areas outside the DPAP.

When the rains fail, agriculture is usually the first to be affected because of its critical dependence on stored soil water. First, soil water in the uplands starts to deplete. Then shortage of water starts to affect people collectively and individually.

The socio-economic impact of drought occurs sooner as frequent droughts have weakened the capacity of the people to bear shock. While drought is basically caused by erratic and deficient rain, the problem has been aggravated by large-scale open-cast mining; deforestation; irregular and non-scientific mining and quarrying; inefficient management of resources, particularly water resources; and decline in traditional systems of water management.

The different strategies for drought management are discussed under the following heads.

- *Adjusting the plant population*: The plant population should be lesser in dryland conditions than under irrigated conditions. The rectangular type of planting pattern should always be followed under dryland conditions. Under dryland conditions whenever moisture stress occurs due to prolonged dry spells, under limited moisture supply the adjustment of plant population can be done.

- *Increasing the inter row distance*: By adjusting more number of plants within the row and increasing the distance between the rows reduces competition during any part of the growing period of the crop. Hence it is more suitable for limited moisture supply conditions.

- *Increasing the intra row distance*: Here the distance between plants is increased by which plants grow luxuriantly from the beginning. There will be competition for moisture during the reproductive period of the crop. Hence it is less advantageous as compared to above under limited moisture supply.

Mid contingent plan for aberrant weather conditions

The contingent management practices done in the standing crop to overcome the unfavorable soil moisture conditions due to prolonged dry spells are known as mid season conditions.

- *Thinning*: This can be done by removing every alternate row or every third row which will save the crop from failure by reducing the competition

- *Spraying*: In crops like groundnut, castor, redgram, etc., during prolonged dry spells the crop can saved by spraying water at weekly intervals or 2 per cent urea at week to 10 days interval.

- *Ratooning*: In crops like sorghum and bajra, ratooning can practiced as mid season correction measure after break of dry spell.

Mulching: It is a practice of spreading any covering material on soil surface to reduce evaporation losses. The mulches will prolong the moisture availability in the soil and save the crop during drought conditions.

142   Rainfed Agriculture

Weed control: Weeds compete with crop for different growth resources ore seriously under dryland conditions. The water requirement of most of the weeds is more than the crop plants. Hence they compete more for soil moisture. Therefore the weed control especially during early stages of crop growth reduces the impact of dry spell by soil moisture conservation.

Water harvesting and life saving irrigation: The collection of runoff water during peak periods of rainfall and storing in different structures is known as water harvesting. The stored water can be used for giving the life saving irrigation during prolonged dry spells.

Short term strategy of contingency planning

Immediate planning:

- Judicious use of surface and groundwater for drinking and irrigation and districts.
- Ensuring availability of quality fodder to animals.
- Livestock management including establishment of fodder/feed depots and cattle camps.
- Selection of crops, cropping sequences and agronomic practices for drought affected areas.
- Promotion of subsidiary income and employment generating activities.
- Gainful implementation of NREGA, RKVY, NFSM, NHM, RGGVY, BRGF and other schemes.
- Deployment of information technologies.

Contingent cropping

- Date bound contingency crop plan for different meteorological sub-divisions and agro-ecological regions.
- Early maturing varieties of different *kharif* crops.
- Crops and cropping system-wise agronomic practices.
- Technology for *rabi*/summer rice in boro cultivation area.
- Potential districts for *rabi* maize.
- Soil and water conservation measures for different rainfall regions.

Promotion of subsidiary income and employment generating activities:

- Extraction of gum from arid land trees and bushes such as *Acacia senegal*.
- Collection of *Prosopis juliflora* pods and its post harvest processing as animal feed and human food.
- Mushroom cultivation, bee keeping, sericulture, tasar cultivation etc.
- Salt making from saline ground water.
- Commercial rising of the nursery for trees, vegetables and annual flowers.
- Multiplication of root stocks as well as nursery of fruits and flowers.

Compensatory production for *kharif* deficit

- Boro rice.
- Winter maize.
- Wheat and other *rabi* crops.
- Intensification of *rabi* pulses and oil seeds.

Medium and long term strategy

- Securing good quality water in drought prone areas.
- Perennial and non-conventional fodder.
- Improved livestocking, breeding and management.
- Up gradation and fine tuning of crops, cropping and farming systems.
- Exploiting underexploited and underutilized plant resources.
- Creation of alternate income and employment generating opportunities.

Major policy issues

- Land related policies.
- Water related policies.

Other policies

- Feed, fodder and seed banks.

- State level policy for livestock.
- Inclusion of coarse cereals in PDS and procurement programme.
- A unique 4-5 years rolling system of credit and repayment.
- Value addition and marketing.
- Strengthening of power distribution grids for hiked demand.
- Augmented supplies of diesels.

Contingent plan for *rabi*

- Time bound *rabi* plan based on receipt of rainfall in August and September.
- Resource conservation technologies including zero tilled specially in Indo-gangetic plains.
- Need based location specific agronomic manipulations.
- Resorting to cultivation of less water demanding crops and cropping systems in limited irrigation water supply areas.

Preparation of appropriate crop plans for dryland areas

Although, the past meteorological records help sufficiently in planning for future, but because, the drought is a phenomenon which may occur at any time during the crop period, so, we have to make our planning accordingly. All India Coordinated Research Project for Dryland Agriculture has evolved several contingent crop production strategy to meet the challenges of acute drought conditions, which are discussed here under –

Modern practices:

- Choice of suitable crops.
- Choice of suitable crop varieties.
- Cropping system (intercropping).
- Planting pattern.
- Planting time.
- Fertilizer management with dose, time and method of application.
- Seed hardening.

(a) Mid season corrections

- Alternate crop strategy
- Transplanting of pearl millet
- "Life-saving" irrigation
- Adjustment of plant population
- Leaf stripping.

(b) Moisture conservation practices

- Use of mulches
- Use of antitranspirants and light reflectants
- Interculture operation
- Weeding
- Water harvesting techniques
- Traditional practices
- Contour bunding
- Ridge and furrow system
- Deep ploughing.

At the household level, the strategies adopted before drought include:

- Crop diversification/ crop variety diversification
- Livelihood diversification
- Staggering of seedling periods
- Early migration
- Keeping land fallow
- Limited resowing.

Strategies adopted after drought sets include:

- Increased resowing if the rains come
- Shift to pulses on upland and medium-level land
- Early sowing of *rabi* crop
- Providing irrigation to paddy seedlings.

146   Rainfed Agriculture

- Transplantation of over-mature paddy seedlings to get straw for feeding animals
- Migration
- Asset depletion
- Borrowings
- Reduction in consumption.

In fragile zones and among the poorest households, reduction in consumption is ultimate coping mechanism.

Appropriate techniques for successful crop production in dryland areas

In every region of the world it is necessary to find or develop appropriate techniques for agriculture. A large part of the surface of the world is arid, characterized as too dry for conventional rainfed agriculture. Yet, millions of people live in such regions, and if current trends in population increase continue, there will soon be millions more. These people must eat, and the wisest course for them is to produce their own food. Yet, the techniques are so varied that only a very large volume would cover the entire subject. This publication is only a primer, an introduction to appropriate techniques. More extensive treatments are mentioned in the bibliography. In many cases the most suitable techniques for a particular region may be those already developed by the local inhabitants. In some cases it will be difficult to improve on local techniques, but at times even simple and inexpensive innovations may be almost revolutionary. This technical note suggests that one must begin to improve local agriculture in arid zones by learning what is already there. Then both techniques and plants that may be useful in specific situations are suggested.

Definitions and degrees of aridity

Arid implies prolonged dryness, and is used with respect to the climate itself, and the land below it. In such regions the ability to produce agricultural crops is restricted. Usually on arid lands the potential evaporation of water from the land exceeds the rainfall. The land may be characterized according to the degree of aridity as dry forest, chaparral or brush land, grassland or savannah, or desert. The word, "arid" does not adequately characterize the soils, however, for they may vary in many ways. Often they are alkaline or saline. Several degrees of dryness must be recognized. The first is where the dry climate is modified by seasonal rainy seasons. In such a region it might be possible to produce a wide range of annual crops during the short rainy season, enough to sustain animals and feed mankind, although few food or feed trees might be feasible without special techniques. The second situation is a year round aridity, sometimes modified by light or irregular rains, which might make production of crops impossible. The third situation is where water is brought in by wells, canals, or other means so that normal agriculture can exist, in spite of the aridity of the climate. This prime concerns the first two situations, but not the third. There are techniques suitable for all arid regions.

Selection of suitable variety

The varieties or hybrids suitable for dry lands should have following characteristics

   i) Short duration, medium height, high yielding ability.

   ii) Big ear head size with bold grains.

   iii) Resistant to water stress conditions.

   iv) Strong penetrating root system.

   v) High harvest index

Seeding time for dryland crops

Proper time of seeding is important in dry lands as the late growing season is likely to be shortened. For this rainfall probabilities eg. cotton, red gram, horse gram dry seeding in 24th meteorologist week at Solapur found optimum. The off season tillage to be practiced to shorten the time aces between first rain and actual seeding time. It also helps to increases moisture.

Timely seeding of *rabi* crops is also important eg. sunflower and safflower - traditional practice - end of September. Improved practice - first fortnight of September. This helps in better utilization of soil moisture and nutrients.

Timely seeding for pest avoidance

Timely seeding of kharif crops found useful in avoiding pest incidence. *e.g. kharif* sorghum should be sown before early July to avoid shoofly incidence at seeding stage and midge fly incidence at flowering to grain formation stage.

Planting pattern and plant densities

Under adequate soil moisture conditions change in planting pattern has no advantage. However, it is necessary while adopting intercropping systems to accommodate intercrop rows.

148 Rainfed Agriculture

e.g. Kharif - bajra + tur in paired planting in 2: 1 row proportion (30 - 15cm). Under limited soil moisture paired planting is useful during the season for efficient moisture use *eg.* rabi sorghum 30 - 30 - 60 cm or 45 - 45 - 90 cm spacing. This is due to deeper and more root growth and convenience in inter culture operations.

Plant density - While deciding the plant density, the availability of stored soil moisture needs to be considered.

Gram - low soil moisture - wider planting - 60 cm. high soil moisture - closer planting - 30 cm.

Sorghum - low soil moisture - 5 - 10 plants / m^2. High soil moisture - 5 -10 plants / m^2.

Safflower - not affected by plant density, bajra - 10 - 15 plants / m^2 optimum. Safflower - 1 to 1.25 lakhs plants / ha optimum.

The optimum plant population leads to higher production per unit area. Optimum plant population of important crops is given below:

Sr.No.	Crop	Spacing (cm)	Plant population in lakhs / ha
1.	Bajra	45 x 15	1.5
2.	Groundnut	30 x 15	2.5
3.	Red gram	60 x 20	0.75
4.	Horse gram	30 x 10	3.30
5.	Moth bean	30 x 10	3.30
6.	Setaria	30 x 5	6.0
7.	Sunflower	60 x 20	1 to 1.25
8.	Gram	30 x 10	3.3

Dry farming practices

Dry farming builds upon knowledge of general agriculture but carries out its practices in the light of the significant probability that this year or next will be a drought. The following agriculture practices are discussed with this background.

Bunding: The first essential step in dry farming is bunding. The land is surveyed and level contours determined every hundred feet. For unusual slopes, it is recommended that for every fall of two feet, a bund 18 to 24 inches in height be constructed. Even when land is fairly flat, a 12 inch height bund at every 250 feet is still found useful. Excess storm water is released by constructing periodic waste weirs with a sill of one-half bund height. This will retain water and minimize the loss of top soil. In order to make the bunds, land must be marked by the surveyor with bund lines. A few feet on either side of it, the land should be ploughed and harrowed. The bund former should be worked along the bund twice, side by side, leaving a furrow in between. This furrow in the middle should be filled in with soil from the ploughed portions on both sides, by means of a scraper. The outlets or "waste weirs" should be constructed of stones. The natural drainage of the area must not be completely stopped but should be controlled by providing suitable outlets for excess storm water to pass gradually, without carrying much silt with it, and after fully saturating the soil and subsoil. The major natural drains in each village area or watershed must be properly maintained so that all fields have some outlet for the extra storm water.

Strip cropping: Strip cropping is a technique that serves to control erosion and increase water absorption thereby maintaining soil fertility and plant response. In effect, it employs several good farming practices such as crop rotation, contour cultivation, stubble mulching, *etc.* By growing in alternating strip crops that permit erosion and exposure of soil and crops that inhibit these actions, several functions are performed.

- Slope length is maintained.

- Movement of runoff water is checked.

- Runoff water is desilted.

- Absorption of rainwater by soil is increased.

- Dense foliage of the erosion resisting crop prevents rain from beating directly on the soil surface.

Strips are, of course, planted perpendicular to either the slope of the land or the prevailing wind direction, according to whether water or wind presents the more serious erosion potential. Additionally, crops which do not resist erosion should be rotated with crops which do. Research has shown that:

- A normal seed rate of groundnut (peanut) is an efficient and suitable crop for checking erosion.

- The normal seed rate of leguminous crops other than groundnut does not provide sufficiently dense canopy to prevent raindrops from beating the soil surface; is should be raised to three times the normal seed rate.

- On the average, the most effective width of contour strips for cereals such as sorghum and millet is 72 feet and for the intervening legume, 24 feet.

150   Rainfed Agriculture

Summer fallow: All of the principles of water conservation and utilization pertaining to dryfarming will not make a crop grow if sufficient rain does not fall. Where the soil depth exceeds 18 inches (450 mm), however, it has been shown that it is possible to store water as soil moisture from one year to the next by the use of proper summer fallow techniques. With a soil depth of 10 to 15 feet, up to 75% of the incident water may be retained though 20 to 40% is more normal. Thus, in an area that averages sufficient rainfall for crop growth, it will be rare that the sum of the stored water and incident water will not be sufficient for crop production. Where families in India have faithfully set aside 5 to 6 acres for summer fallow each year, drought-induced famine has been virtually eliminated. The partial loss of a crop in the year of fallow is offset to a great extent by a very much increased yield in the year of cropping. Such increased yield in a year of failure of the general crop in the surrounding areas has a far greater value than a normal crop of a good season. In order to accomplish this objective, the soil must be loose and permeable to soak up the rainfall and the dirt/stubble and mulch must be maintained to minimize evaporation. The land is worked with a tine-cultivator followed by occasional harrowing, particularly after rainfall, and weeds (which use as much or more water as crops) must not be allowed to grow. Though this expenditure on cultivation is relatively small, neglecting to provide the surface mulch at any time may cause more moisture to evaporate in a few, hot days than would fall during the whole season. Experience has shown that where rainfall is 10 to 15 inches per year (250 to 375 mm/yr.) a clear fallow every other year is necessary and, at 15 to 20 inches per year (375 to 500 mm/yr.), every third year.

Mulches: Water easily enters porous soil and, as it seeps downward, becomes absorbed as films of water around the soil grains. These films form a continuous column of water to the surface of the soil. The film tends to remain the same thickness around all the soil grains with which it is in contact. This film of water in the soil is known as the capillary water and is the source of water for the plants. The sun, wind, and dry air will cause evaporation at the surface, thus reducing the thickness of the film at the surface. The thicker films in the subsoil will rise to equalize the distribution again. This will continue until the films are so thin that the plant roots can draw no further moisture from them. The result is drought.

Stubble mulching: Aims at disrupting the soil drying process by protecting the soil surface at all times, either with a growing crop or with crop residues left on the surface during fallows. To be effective, at least one ton per hectare must cover the surface, and the maximum benefit per unit residue is obtained at about two tons per hectare. Benefit may still be obtained at 8 tons per hectare.

The first benefit of stubble mulch is that wind speed is reduced at the surface by up to 99%, significantly reducing losses by evaporation. In addition, crop and weed residues can improve water penetration and decrease water runoff losses by a factor of 2 to 6 times and reduce wind and water erosion by factors of 4 to 8 relative to a bare fallow field. There are two limitations to the advantages of stubble mulch farming:

1. Dead surface vegetative matter can provide a home/breeding ground for plant diseases, insects or rodents. Use of mulch not related to the succeeding crops will minimize much of the disease and insect effects. Use of stubble mulch only in the dry season will minimize all biological activity.

2. For decomposition, the ideal carbon to nitrogen ratio (C/N) is 25 to 30. Dry, woody, or non-green straw, stalks, etc. have a C/N of 50 to 100. This tends to slow decomposition and deplete soil nitrogen temporarily.

Nitrogen is a major requirement for protein synthesis by plants. A stubble mulch during a biologically active period such as the rainy season, should only be used when either:

- Soil nitrogen is very high.

- Plant nitrogen needs are very low (such as cassava).

- A nitrogen-containing fertilizer is used.

To obtain the benefit of mulching on soil structure without causing temporary de-nitrification, the mulch can be composted before adding it to the soil. Rapid bacterial action in the tropics makes composting less beneficial than in temperate climates but may still be worthwhile.

Dirt mulching: Aims at disrupting the soil drying process with tillage techniques that separate the upper layer of the soil from the lower layers, making the soil moisture film discontinuous. In addition the soil surface is made more receptive to water intake. Principles of dirt mulching:

- Effectiveness increases with increasing depth to a limit of to 4 inches (75 to 100 mm).

- Increasing the dirt mulch depth decreases the available fertile soil.

- The effectiveness of dirt mulches decrease with age. Consequently it must be recreated by shallow tillage of harrowing after each rain or each month (whichever is more frequent).

- The crumb form of dirt mulch (particles greater than 1 mm) is more effective and resists wind erosion more than the dust form.

152   Rainfed Agriculture

- Dirt mulches can only be properly made when the soil is moist.

- For a climate with a "rainy" growing season and a hot, windy, dry season, dirt mulching should only be performed during the rainy season and with a growing crop to slow the wind and water and hold the soil. The improper use of dirt mulch presents serious erosion potential. The "dust bowl" condition in the Great Plains of the U.S. that destroyed or damaged millions of acres of prime cropland was a direct consequence of the use of the dirt mulch.

Tillage practices: Ploughing, when the soil is in the proper condition, wears the soil into thin layers, and forces the layers past each other. If the soil is too wet when ploughed (especially if it is heavy), the soil crumbs or granules are destroyed, thus puddling or compacting the soil. When the soil is too dry, the soil tends to pulverize and form dust. Ploughs with steep mouldboards have the greatest pulverizing action upon the soil. The plough with the less steep mouldboard has fewer tendencies to puddle the soil and is of less draft.

- To produce a rough, cloddy surface that will increase moisture absorption and reduce runoff, as well as erosion from wind and water.

- To control/destroy weeds that competes with crop for sunlight, nutrients, and water.

- To destroy or prevent the formation of a hard pan which can develop after repeated shallow ploughing or harrowing. This hard pan can stunt root growth, reduce water storage, and check the capillary rise of water from the subsoil.

- Promote bacterial activity by aerating soil, encouraging the decay of residues and the release of nutrients.

Timing of tillage: Ploughing, like planting, is sensitive to moisture and neither should be done when soil is either too wet or dry. In the arid and semiarid tropics, proper moisture conditions are likely to occur only at the beginning of the rainy season and should be done on the same day. If possible, planting should immediately follow ploughing, with seed rows centred on the furrow slices. A crosswise harrowing will cover seeds and close air spaces, thus creating dirt mulch and keeping out the drying winds. If the crop is then harrowed/cultivated several times during the season, especially after rains, much moisture will be conserved. The proper soil moisture condition for ploughing is indicated by a manual soil test. The usual test is to squeeze a handful of soil. If it sticks together in a ball and does not readily crumble under slight pressure by the thumb and finger, it is too wet for ploughing or working. If it does not stick in a ball, it is too dry. When examining soils, samples should be taken both at and a few inches below the surface. Soil that sticks to the plough or to other tools is usually too wet. A shiny, unbroken surface of the turned furrow is another indication of excessive soil moisture. In general, sandy soils and those containing high proportions of organic matter bear ploughing and working at higher moisture contents than do heavy clay soils. In semi-arid regions, the soil after harvest time is generally too dry for good ploughing. Yet if the field is left uncultivated, this dry condition may become even worse and weeds will also grow and go to seed. The field should be harrowed (or ploughed without mouldboard) and crop residues left to form a stubble mulch to absorb/retain moisture and soil until the rains return. Stubble should not be immediately covered and incorporated in the soil unless rodent or insect infestation is heavy (and even then burning should be considered). It has been well demonstrated that it is normally impossible to raise the soil organic matter content in areas where temperatures are high for long periods. When moisture is present, the rates of oxidation are extremely high and incorporated organic matter is lost quickly. The benefits thus derived from decomposition, as occurs in the more temperate regions, are not normally experienced. When left on the surface, however, organic matter does not decay so rapidly. Incorporation with the soils will tend to depress the levels of available nitrogen, to the detriment of crops if soil nitrogen is low. If soil nitrogen levels are adequate, the incorporation of residues to the soil may be beneficial if done with spring ploughing at the start of the rainy season.

Depth of ploughing: Generally speaking, heavy clay soils should be ploughed deeper than light, sandy soils, in order to promote circulation of the air and bacterial activity. Deep ploughing on sandy soils, which are naturally porous and open, tends to disconnect the seed bed from the subsoil and speeds soil drying by too free a circulation of air in the soil.

In semi-arid climates, the greatest advantage to be gained from deep ploughing (5-8 inches) is the development of a comparatively large moisture reservoir. When land is not ploughed more than 3 or 4 inches deep for a period of years, a hard plough sole is very likely to form, through which roots and rain can only penetrate with difficulty. A shallow plough sole will saturate quickly with rainwater and increase runoff rates. As a rule, tillage below 5-6 inches also causes increased evaporation rates, losing precious water. This deep ploughing need not necessarily be done annually. Depending on soil and rainfall, a deep ploughing of 5-6 inches every 2 to 5 years is satisfactory. As noted earlier, the soil mulch attains maximum effectiveness at a depth of 3-4 inches which can be maintained with a hand harrow/cultivator.

154 Rainfed Agriculture

Deep ploughing in some clay and loam soils will reduce yields for one or two seasons afterward as a result of acidic subsoil. This may be dealt with by liming the soil (neutralizing the acidity) or by varying the depth of the ploughing slowly so that the acidic subsoil is exposed a little at a time. This practice will also eliminate the plough sole.

Intertillage / cultivation: Crops sown in rows can take advantage of intertillage practices which serve three basic functions:

- Easy weeding without meticulous hand labour. Weeds compete for moisture and nutrients, thus they should be destroyed while small, before they have grown more than 2 or 3 leaves. If seeds are broadcast, or thickly sown, they can at best only be cultivated manually, a back-breaking task.

- Increase the formation of nitrates by bacteria. Cultivation aerates the soil and forms a mulch of dead weeds and stubble on which bacteria operate and form nitrates. Cultivation for this purpose should be undertaken during the early period of plant growth, and should be relatively deep, on the order of 2-3 inches.

- Intertillage conserves moisture by the formation of a dirt mulch as described earlier. It is imperative that cultivation be performed after rainfalls. Even a light rain can re-form capillary connections between the stored soil moisture and the surface of the ground. After a few drying days like that, it is possible for soil moisture to be lower than before the rainfall.

Crop rotation: One of the first principles of dry farming with regard to cropping practices is that crop rotation as practiced in more humid regions is not necessarily recommended in semiarid lands. The following constitute the chief differences:

- Only a limited number of crops are adapted to the climatic conditions and the farmer must sow the crop best suited to the moisture conditions encountered at that time.

- Moisture is so dominantly limiting, that "soil improving" crops are much less effective than in more humid areas.

- Success with rigid or complex sequences is difficult in the face of widely varying rainfall.

Crop substitution: Alternate crop strategies have been worked out for important regions of the country for *vertisols, alfisols, entisols, submontane* and *sierozemic* soils. Strategy has also been evolved for normal onset of rains, breaks in rains, early withdrawal, its uneven distribution; through selection of uneven crops/varieties which are inefficient utilize of the soil moisture, less responsive to production input and potentially low producers should be substituted by more efficient ones. Appropriate crops, suits varying rainfall situations, have been identified for most of the dry land regions of India.

Crops which do not grow under normal rainfall years may not do so under abnormal years. Studies conducted in agro climatic conditions of Varanasi (eastern U.P.) revealed that under normal monsoon crops like short duration upland rice, maize, pearl millet, blackgram, greengram, sesame, pigeonpea *etc.* should be taken up on the basis of needs. These crops should be followed by chickpea, lentil, barley, mustard, safflower, linseed *etc.* on residual moisture during winter season.

If monsoon sets in as late as second week of July, short duration upland rice (variety - NDR-97 and NDR-118) may be included in place of Akashi and Cauvery is recommended. If the rains are delayed beyond the period but start somewhere in last week of July or first week of August and growing season is reduced to 6-7 days, then cultivation of hybrid pearl millet (NHB 3-4, B.J. 104), blackgram (Type 9), greengram (Jagriti and Jyoti) may be included in pace of T-44 and K-851 etc. should be grown. Yet another alternative could be to harvest a fodder of either pearl millet, maize, sorghum or a mixture of cowpea, blackgram and one of the above fodder crops.

In case monsoon rains stop early towards the end of season, normal sowing of short duration upland rice, black gram and sesame may be taken up. If the rain stops very early, i.e. by the end of August or first week of September, only fodder crops or grain legumes could be harvested. Depending upon the soil moisture condition, relay sowing of crops like chickpea, lentil, mustard, linseed and barley could be done in rabi season.

Traditional and alternate efficient crops in different dryland regions of India

S.No.	Region	Traditional crop	Alternate efficient crop
1.	Deccan plateau	Cotton, wheat	Safflower
2.	Malwa Plateau	Wheat	Safflower, chickpea
3.	Uplands of Bihar plateau and Orissa	Rice	Ragi, blackgram, groundnut
4.	South-east Rajasthan	Maize	Sorghum
5.	North Madhya Pradesh	Maize	Soybean
6.	Eastern UP	Wheat	Chickpea
7.	Sierozems of North-west India	Wheat	Mustard, taramira (*Eruca sativa*)

During the recent drought, it was found that farmers in Karnataka, Andhra Pradesh and Maharashtra who went in for sunflower cultivation were in gainers. Sunflower succeeded where other crops failed. In other dry land regions, alternative efficient crops can profitably substitute the traditional ones.

Relative yield of traditional and efficient crops in dryland areas

Region	Traditional	Yield (q/ha)	Efficient crops	Yield (q/ha)
Bellary	Cotton	2.0	Sorghum	26.7
Varanasi	Wheat	8.6	Chickpea	28.5
Ranchi	Upland rice	28.8	Maize	33.6
Indore	Greengram, Wheat	11.8	Soybean	33.3
		11.0	Safflower	24.2
Agra	Wheat	10.3	Mustard	20.4
Hisar	Wheat	3.0	Taramira	16.0
Udaipur	Maize	18.0	Hybrid sorghum	29.0

Dry land research has remained confined to important traditional crops such as sorghum, millet, pulse and oilseeds and has not explored the possibility of growing non-traditional crops such as dye-providing crops. *Eg.* Henna (*Lawsonia inermis*: mehadi) and jaffra(*Bixa ovellana*) species (*eg.* cumin), and medicinal value crops (*eg.* citronella, lemon grass, senna and isabgol). These crops need to find an important place in research agenda of dry land farming.

Time has come for the relevant researchers to plan a joint integrated research programme for maximizing the profitability, productivity and sustainability of learning systems of rainfed areas. Sericulture offers great promise in rainfed farming strategy, particularly of the watershed approach in peninsular India.

Efficient cropping system: Besides putting various measures to increase the productivity levels of dry land crops, efforts would also be needed to increase the cropping intensity in dry land areas which was generally 100%, implying that a single crop was taken during the year. Cropping intensities of these areas could be increased by practice of inter cropping and multi cropping (sequential) by way of more efficient utilization of resources. The cropping intensity would depend on the length of growing season which in turn depends on rainfall pattern and the soil moisture storage capacity of the soil. For example in Indore region, receiving 1000 mm annual rainfall, only single crop can be taken on shallow soils, inter cropping in medium depth soils and double cropping on deep soils. Similar crop combinations have been identified for different regions of the country. In dry land of Varanasi region upland rice-chickpea/lentil sequence can be practices with advantage.

Inter cropping of vegetables with grain crops was pursued vigorously in some centres such as Varanasi and Phulbani. At both the places long duration pigeonpea was inter cropped with vegetables such as okra, radish and chilli. Such inter cropping systems would be very useful to get maximum returns from rainfall agriculture. Even at Solapur, leafy vegetables and some short duration beans were grown as intercrops during the rainy season.

Alternate land use system: All dry lands are not suitable for crop production. Some lands may be suitable for range/pasture management, while others for tree farming, ley farming, dry land horticulture, agro-forestry systems including alley cropping. All these systems which are alternatives to crop production are called as alternate land use systems. This system not only helps in generating much needed off-season employment in mono crop drylands but also minimizes risk, utilizes off season rains which may otherwise go waste as runoff, prevents degradation of soils and restores balance in the ecosystem.

Crop production may be disastrous in the years of drought, whereas drought resistant grasses and trees could be remunerative. Scientists of dryland have developed many alternate land use systems which may suit different agro ecological situations. These are alley cropping, agri-horticultural system and silvi-pastoral systems which utilize the resources in better way for increased and stabilized production from dry lands.

i. **Alley cropping**: For imparting stability and providing sustainability to the farming system, a tree-cum-crop system will be one most appropriate for such situations. One such system called 'alley cropping' - a version of agro-forestry system, could meet the multiple requirements of food, fodder, fuel, fertilizer etc. Alley cropping is a system in which food crops are grown in alleys formed by hedge rows of trees or shrubs. The essential feature of the system is that hedge rows are cut back at planting and kept pruned during cropping to prevent shading and to reduce competition with food crops. For example, fast growing leguminous trees such as *Leucaena leucocephala* or *liliricidia* spp. are planted in rows. During the cropping season, trees are lopped at about 0.5 metre height. These loppings are used as much to reduce moisture loss and improve the nutrient status of soil. Arable crops like maize, rice, pearl millet, legumes, oilseeds etc. are planted in the alleys formed by the two rows of threes. This is also known as agri-silviculture. Alley cropping is also a form of conservation farming which enhances soil fertility and prevents erosion. One very strong argument in favour of alley cropping is its ability to produce usable material even in years of severe drought. At Rajkot in 1985, rainfall received during the season was only 30% of the normal. There was total failure of grain

158   Rainfed Agriculture

production of the three legume crops tried in the system. In sole crop plots production was limited to 5.0 q/ha to 17.0 q/ha of green fodder. However, in alley cropped plots, *Leucaena* hedge-rows produced over 50.0 q/ha of green fodder.

ii. **Agri-horticultural system**: Agri-horticultural system plays an important role in dry land areas, especially in semi-arid regions where production of annual crops is not only low but also highly unstable. Fruit trees if suitably integrated in dryland farming system could add significantly to overall agricultural production including food, fuel and fodder, conservation of soil and water and stability in production and income. Dryland fruit trees being deep rooted and hardy, can better tolerated monsoonal aberrations than short duration seasonal crops. Hence, in drought year when annual crops usually fail or their production is highly depressed; fruit trees species yield considerable food, fodder and fuel. A suitable example of agri-horti-system is growing of cow pea/green gram/horse gram in inter space of *ber (Zizyphus mauritiaria)* at 6 x 6 m spacing at Hyderabad. Phalsa (*Grewia asiatica*) may be planted in between two ber plants in a row with a view to intensify the system. A well managed dry land orchard of *ber* should give 50 kg fruits per tree/year. There should be 250 plants/ha for optimized production. The grow income would touch around Rs. 50,000/ha (250 x 50 x 4), assuming that one kg *ber* fetches Rs. 4. One could get an additional income of Rs. 800 Rs. 1000 from green gram/cow pea (2.5-3.0 q/ha).

iii. **Silvi-pastoral system**: This system is suited to marginal dry lands and is most preferable where the fodder shortages are experienced frequently. The system essentially consists of a top feed tree species carrying grasses on legumes (preferable perennial) as understorey crops. Dry land farmers having larger holdings and keeping a land follow for a longer period for one reason on the other should go in for this system which could provide both fodder and fuel. In a survey carried out in Andhra Pradesh, Karnataka and Maharashtra by CRIDA scientists, it was revealed that after food it is the fodder which is of paramount importance for sustaining animal wealth in rural areas. In years to come, fuel will assume greater importance. In August, 1981 *Leucaena leucocephalla* was planted in contour trenches 7.5 m apart, the plant to plant spacing being maintained at 2.0 m at CRIDA. Four strips at upper reaches of plot (2% slope) were put under *Cenchrus ciliaris,* while lower four strips were seeded with *Stylosanthes hamata.* The system has come up very well.

Efficient implements: In order to take full advantage of annual precipitation in dryland agriculture, higher doses of energy input is essential. Farmers in dry

Drought and Its Management 159

lands have been using traditional and outdated farm equipments which not only perform poorly but also demand a lot of energy and time and post-harvest operations. Farm implements can help to conserve as much rain water *in-situ* as possible and to harvest rain water. Shallow off season tillage with pre-monsoon showers ensures better moisture conservation and lesser weed intensity. It has resulted in 20% yield increase in sorghum in Andhra Pradesh. Deep tillage helps in increasing water in soils having textural profiles and hard pan. This has resulted in 10% yield increase in sorghum and 9% yield increase in case of caster. For *in-situ* moisture conservation, land has to be opened so that it can cause hurdle to flow of rain water. Tillage machines of appropriate size and type matching the power sources need to be used. Location specific seeders have been developed for dry land areas and these have shown good prospects and promise. A feature of these machines is that the seeds and fertilizers are placed in the moist zone of the soil resulting in a high percentage of seed germination and good crop vigour. In deciding farm mechanization in dry land areas, where farmers are generally poor, and their socio-economic condition should always be kept in mind.

The foregoing discussions show that technology of crop production in dry land areas have been generated to a great extent. What is important now is to view it in socio-economic context of the farmers. Once the technology is adopted by the farmers, the contribution of dry land areas to the total production can be sizably improved and the living standards of the farmers of these areas can be improved. This has been clearly shown in selected watershed areas and what is needed is to have more watersheds identified, proper technology to be developed and implemented.

8

Land Use Capability Classification, Scope of Agro-Horticultural, Agro-Forestry and Silvi-Pasture in Dryland Agriculture

Land capability classification

Introduction: The United States Department of Agriculture (USDA) land classification system is interpretative, using the USDA soil survey map as a basis and classifying the individual soil map units in groups that have similar management requirements. At the highest of categorization, eight soil classes are distinguished, namely, arable lands (I to IV) and non arable lands (V to VIII).

Origin of land capability classification (1933) : The Georgia Piedmont Land Capability Classification (LCC): First U.S. effort to systematically determine the best use of lands by classifying and mapping erosion rates and potential in relation to both physical characteristics and agricultural capacity.

Definition: Land capability classification is a system of grouping soils primarily on the basis of their capability to produce common cultivated crops and pasture plants without deteriorating environment over a long period of time.

Classes: Land capability classification is subdivided into capability class and capability subclass nationally. Some states also use a capability unit.

The capability classification provides three major categories of soil groupings:

(1) Capability unit, (2) capability subclass, and (3) capability class.

Capability unit: A capability unit is a grouping of one or more individual soil mapping units having similar potentials and continuing limitations or hazards.

The soils in capability unit are sufficiently uniform to (a) produce similar kind of cultivated crops and pasture plants with similar management practices, (b) require similar conservation treatment and management under the same kind and condition of vegetative cover, (c) have comparable potential productivity. The capability unit condenses and simplifies soils information for planning individual tracts of land, field by field. Capability units with the class and subclass furnish information about the degree of limitation, kind of conservation problems and the management practices needed.

Capability subclass: Subclasses are groups of capability units which have the same major conservation problem, such as; e-Erosion and runoff; w-Excess water; s-Root-zone limitations; c-Climatic limitations. The capability subclass provides information as to the kind of conservation problem or limitations involved. The class and subclass together provide the map user information about both the degree of limitation and kind of problem involved for broad program planning, conservation need studies, and similar purposes.

Capability class: Capability classes are groups of capability subclasses or capability units that have the same relative degree of hazard or limitation. The risks of soil damage or limitation in use become progressively greater from class I to class VIII. The capability classes are useful as a means of introducing the map user to tile more detailed information on the soil map. The classes show the location, amount, and general suitability of the soils for agricultural use. Only information concerning general agricultural limitations in soil use are obtained at the capability class level.

Aim of land capability classification

- LCC classes were designed to identify arable land and to inform farm planners and farmers of the maximum level of sustainable tillage intensity that could be applied.

- Best land use is determined by how the land will give the most benefits to people.

- Suitability of land for agricultural uses.

- Usage should not cause damage to the land although nutrients may be removed.

Land capability classes

Use	I	II	III	IV	V	VI	VII	VIII
Row crops	X	X	/					
Hay, small grains	X	X	X	/				
Pasture	X	X	X	X	X	X		
Range	X	X	X	X	X	X	/	
Woodland	X	X	X	X	X	X	/	
Recreation, wildlife	X	X	X	X	X	X	X	X

Land capability classes according to arable and non-arable land areas by figure.

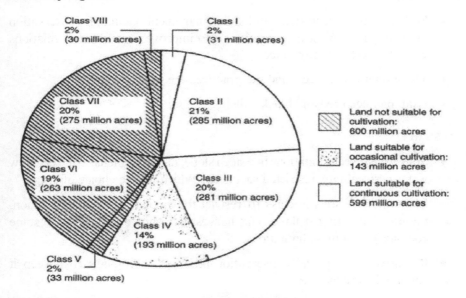

Land capability classes – Arable lands (I-IV)

Class I :

- It has no limitation hence well suited for intensive crop cultivation
- Such lands need only crop management practices like fertilizer, manures, crop rotation *etc.* to maintain their productivity.
- Can be used continuously for intensive crop production with good farming practices
- Soils in this class are suited to a wide range of plants and may be used

164 Rainfed Agriculture

safely for cultivated crops, pasture, range, woodland, and wildlife.

- Example of class I soil are alluvial soil of Indo-gangatic plains

Class II :

- Soils have moderate limitations, reducing cropping options and/or requiring moderate conservation practice implementation
- More limitations than Class I land for crop production.
- 2-5% slope is main difference.
- Such limitation such as gentle slope, moderate erosion hazard, soil structure less ideal, slight to moderate alkali or saline condition somewhat restricted drainage.
- Soils in class II require careful soil management, including conservation practices, to prevent deterioration or to improve air and water relations when the soils are cultivated.
- The limitations are few and the practices are easy to apply.
- Example: deep red and black soils.

Class III :

- Soils have severe limitations like steep slope, high erosion hazard, very slow water permeability, restricted root zone. Which reduce choice of crops.
- Limitations of soils in class III restrict the amount of clean cultivation, timing of planting, tillage, and harvesting; choice of crops; or some combination of these limitations.
- Requires more special conservation practices than Class II to keep it continually productive.
- Slopes 6-10%
- Terraces and strip-cropping are more often used.
- Example; shallow red soils, slightly saline black soil.

Class IV :

- Soils in class IV have very severe limitations that restrict the choice of plants, require very careful management/ or both.
- Same limitations as Class V soils, except more need for sustainable management practices

- Slope 18-30%

- Soils in class IV may be well suited to only two or three of the common crops or the harvest produced may be low in relation to inputs over a long period of time.

- Many sloping soils in class IV in humid areas are suited to occasional but not regular cultivation.

- Some soils in class IV are well suited to one or more of the special crops, such as fruits and ornamental trees and shrubs, but this suitability itself is not sufficient to place soil in class IV.

- Use for cultivated crops is limited as a result of the effects of one or more permanent features such as (1) steep slopes, (2)severe susceptibility to water or wind erosion, (3) severe effects of past erosion, (4) shallow soils, (5) low moisture-holding capacity, (6) frequent overflows accompanied by severe crop damage, (7) excessive wetness with continuing hazard of water logging after drainage, (8) severe salinity or sodium,

- Example; shallow soils, saline soils, alkaline soils

Land capability classes – Non arable lands (V-VIII)

Land limited in use—generally not suited to cultivation

Class V :

- Soils with major physical limitations (other than erosion), making them limited in their use. May be appropriate for forestry, rangeland, or viticulture.

- Soils in class V have little or no erosion hazard but have other limitations impractical to remove that limit their use largely to pasture, range, woodland, or wildlife food and cover.

- Can be used for pasture crops and cattle grazing, hay crops or tree farming

- Class V has some combination of these limitations. Examples of class V are (1) soils of the bottom lands subject to frequent overflow that prevents the normal production of cultivated crops, (2) nearly level soils with a growing season that prevents the normal production of cultivated crops, (3) level or nearly level stony or rocky soils, and (4) pounded areas where drainage for cultivated crops is not feasible but where soils are suitable for grasses or trees.

- These limitations cultivation of the common crops is not feasible but pastures can be improved and benefits from proper management can be expected.

Class VI :

1. Soils in class VI have severe limitations that make them generally unsuited to cultivation and limit their use largely to pasture or range, woodland, or wildlife food and cover.

2. Slope 18-30% or more slopes.

3. Same limitations as Class V soils, except more need for sustainable management practice.

4. Soils in class VI have continuing limitations that cannot be corrected, such as (1) steep slope, (2) severe erosion hazard, (3) effects of past erosion, (4) stoniness, (5) shallow rooting zone,(6) excessive wetness or overflow, (7) low-moisture capacity, (8) salinity or sodium, or (9) severe climate. Because of one or more of these limitations these soils are not generally suited to cultivated crops. But they may be used for pasture, range, woodland, or wildlife cover or for some combination of these.

5. Some of the soils in this class are also adapted to special crops such as sodded orchards, blueberries, or the like, requiring soil conditions unlike those demanded by the common crops.

Class VII :

1. Soil limitations so severe they are unsuitable for cultivation and only suitable for rangeland or forestry.

2. Slope >30%

3. Large rock surfaces and boulders may be found Physical conditions of soils in class VII are such that it is impractical to apply such pasture or range improvements as seeding, liming, fertilizing, and water control with contour furrows, ditches, diversions, or water spreaders.

4. Soil restrictions are more severe than those in class VI because of one or more continuing limitations that cannot be corrected, such as (1) very steep slopes, (2) erosion, (3) shallow soil, (4) stones, (5) wet soil, (6) salts or sodium, (7) unfavorable climate, or (8) other limitations that make them unsuited to common cultivated crops. They can be used safely for grazing or woodland or wildlife food and cover or for some combination of these under proper management.

Class VIII :

1. Soil limitations nearly preclude production; often used for recreation.
2. Soils and landforms in class VIII cannot be expected to return significant on-site benefits from management for crops, grasses, or trees, although benefits from wildlife use, watershed protection, or recreation may be possible.
3. Limitations that cannot be corrected may result from the effects of one or more of the following: (1) Erosion or erosion hazard, (2) severe climate,(3) wet soil, (4) stones, (5) low-moisture capacity, and (6) salinity or sodium.
4. Badlands, rocky outcrop, sandy beaches, river wash, mine tailings, and other nearly barren lands are included in class VIII.

Example; sandy beaches, river washes *etc.*

Soil classification was expanded to identify appropriate uses of non-arable land, including grazing

Second major change: Capability subclasses

Subclass designations identify the character of the soil's limitation. Class I soils are not subdivided.

Subclasses :

e: erosion hazard or damage

w: water in or on the soil interferes with arability

s: root zone limitation: shallow, stony, saline, or droughty

c: climatic limitations, usually very cold or very hot.

Strengths and limitations

- LCC was created as a conservation planning tool, with the goal of sustainability, or "agricultural permanence"
- LCC does NOT map erosion or highly erodible soils.
- Application of LCC as an erosion indicator is NOT appropriate.
- Subclass "e" may indicate *erosion risk* or *past erosion;* in other words, IIIe soils may be more eroded than IVe soils.

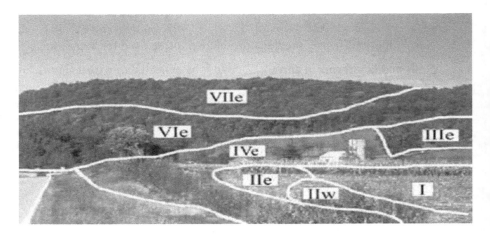

Alternate land use systems

Alternate land use systems are applicable to all classes of land aiming at generation of assured income with minimum risk through efficient utilization of available resources. Agroforestry is a part of alternate land use systems. Agroforestry essentially involves a woody perennial, while alternate land use systems sometimes lack a woody component. Agroforestry is a combination of two or more components like tree with grass, but not tree or grass alone. Alternate land use systems can be defined as a perennial systems / practice adopted to replace or modify the traditional land use. Alternate land use systems possess more potentiality and flexibility in land use than the traditional crop production systems. Alternate land use systems aim at a suitable farming system matching the land capability class like tree farming, ley farming, pasture management etc. In countries like India where land is scarce and labour is in plenty but less productive, coupled with scarce capital and high interest rates, alternate land use systems are an ideal option. There is predominance of marginal and small farmers representing 75 percent of the total holdings of less than 2.0 ha. Small land holding have some inherent constraints to increased productivity, the principal being lack of resources and impoverished soils. Alternate land use systems are more appropriate in areas where subsistence farming is practiced in fragile ecosystems. These systems help in efficient utilization of resources like land and labour (Fig 1).

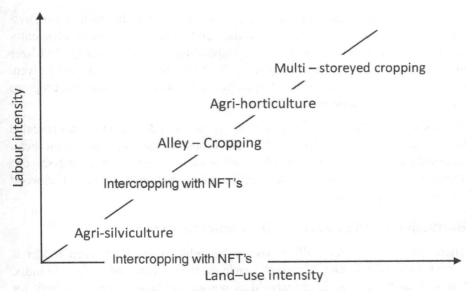

Fig. 1 : Agroforestry systems in relation to labour and land-use intensity.

Agri-horticulture system for non-arable lands

This system is also termed as food-cum-fruit system in which short duration arable crops are raised in the interspaces of fruit trees. Most of the fruit plants developed full canopy after several years and some of them require regular pruning, thus permitting an intercrop. Intercropping provides weed-free area and yield obtained will be a bonus to the farmer. This systems mainly focuses on higher returns per unit area. The economic aspect of fruit culture is also no less important. Well maintained and established orchards bring better returns than field crops from the same piece of land. Fruit growing is a longterm venture and requires a high initial investment. The concept is to encourage small farmers to take up tree planting and ensure good returns through intercropping. This system also helps to generate more employment for the farmer especially during the offseason when the crops are not cultivated. It is very well recognized now that fruits should not be considered a luxury, since they are protective foods necessary for the maintenance of human health. Many fruits are also known to possess specific medicinal values. This is more important for a country like India where the majority of the population is vegetarian.

The system works best in medium to deep soils with good water holding capacity. Individual farm ponds and supplemental watering in the offseason will certainly improve the scope of fruit farming in drylands, if suitable species/varieties are chosen, Jamun (*Syzigium cuminii*), custard apple, guava, mango, *Sapota* are

170 Rainfed Agriculture

the suitable fruit species. The trees can be established in the contour furrows/ saucer shaped pits. The rainwater in the catchment could thus be efficiently used by concentrating it in the pits. Depending on the intercrop, fertilizer application (40-30-0 for cereals and 10-30-0 for pulse crops) has to be given. Generally, it is not desirable to interplant cereals and other exhaustive crops so as to avoid competition with fruit plants.

Studies involving Ber (*Ziziphus mauritiana*) and short duration legumes revealed higher returns than their respective sole crops. Horsegram and clusterbean gave and additional gross return of Rs.3,000/- and Rs.1,900/- per ha, respectively. The gross return from *Ber* was Rs.7,320/- per ha, besides 1.2 tons of fuel wood obtained from pruning.

Horti/Silvi-pastoral systems for non-arable land

There are four hundred million bovines in India, scarcity of green fodder is more felt than dry fodder. Livestock rearing is the mainstay of farmers in India. Class IV and above soils are termed as non-arable lands. These are unfit for crop production. The topography of these lands is highly undulating. They lack vegetative cover, Runoff and soils loss is inevitable. The carrying capacity of these lands is less than 2 sheeps/ha. These lands have to be brought under close canopy vegetative cover to prevent these lands from further degradation, improve the carrying capacity of native grasslands, and for stabilizing the economy of farming community.

Grasses and legumes associated with hardy fruit trees/top-feed trees is the panacea for non-arable lands. Horti-pastoral system is defined as, "One of the agroforestry system which involves integration of fruit trees with pasture. When a fruit tree is replaced with a top-feed tree, it is called as silvi-pastoral system".

Horti-pastoral system

In a horti-pastoral system, fruit trees Guava (*Psidium guajava*) and custard apple (*Annona squamosa*) were planted and seeds of *Stylosanthes hamata* and *Cenchrus ciliaris* were broadcasted one year later to ensure good establishment of fruit trees. After 18 months of pasture establishment survival and growth of fruit was found to be poor with grass than legume association. In terms of herbage yield, *Cenchrus* grass outyielded the stylo legume but fruit plants suffered severely due to moisture stress.

Silvi-pastoral system

Three-month old seedling of *Leucaena leucocephala* were planted in contour furrows in 1981 at 7.5 m apart with plant to plant spacing of 2.0 m on a plot having about 2 per cent slope. As undergrowth, improved pasture species *Stylosanthes hamata* in lower four strips and *Cenchrus ciliaris* in upper four strips were sown.

The performance of *Leucaena* + *Stylosanthes hamata* was found better (Stylo 4.2 t/ha/year DM) than *Leucaena* + *Cenchrus ciliaris* (Cenchrus 2.5 t/ha/year DM). The traditional sorghum crop in a similar soil (0-20 cm soil depth) seldom yielded a biomass of 2 t/ha/year. *Leucaena* trees were felled after eight years (700 trees/ha) and yielded 35 t/ha of biomass. The yield per tree, on an average, was 51.1 kg, of which stem, branch and twigs contributed 60, 24 and 16 percent, respectively.

Silvi-pastoral management system

This system is suited to marginal drylands and is most preferable where the fodder shortage looms large. The system essentially consists of a top-feed tree species carrying grasses or legumes (preferably perennial) as understorey crops. Dryland farmers having larger holdings and keeping a part fallow for a longer period for one reason or the other, should go in for this system which could provide both fodder and fuel. In a survey carried out in Andhra Pradesh, Karnataka and Maharashtra by CRIDA scientists, it was revealed that after food, it is the fodder that is of paramount importance for sustaining animal wealth in rural areas. In years to come, fuel will assume greater importance.

172   Rainfed Agriculture

Farming systems options for different agro-climatic conditions

Annual rainfall (mm)	Soil type	Farming systems	Suitable tree/ grass/ legume species
Less than 500	Shallow (0-30 cm)	Tree farming	*Prosopis cineraria, P. juliflora, Acacia aneuro, A. nilotica, A. tortilis, Pithecellobium dulce*
	Medium (0-45 cm)	Pasture management	*Lasiurus sindicus* (light textured soils). *Cenchrus setigerus, Sehima nervosum, Stylosanthes scabra, Clitoria ternalea*
500-750	Shallow (0-30 cm)	Silvi-pastoral system	*Acacua nilotica, Colophospermum mopane, Dalbergia sissoo, Hardwickia binata, Cassia sturti, Albizia amara, Leucaena, sp. Cenchrus ciliaris, C. setigerus, Dicanthium annulatum, Panicum antidotale, Stylosanthes hamata, Macroptilium atropurpureum*
	Medium (0-45 cm)	Horti-pastoral system	Custard apple, *Bet, Jamun Aonla,* Tamarind, Woodapple, *Bael, Cenchrusciliaris, Panicum antidotale, Urochloa mosambicensis, Stylosanthes hamata, Macroptilium atropurpureum, Clitoria ternatea.*
More than 750	Shallow (0-30 cm)	Ley farming or silvi-pastoral system	3 years *Stylosanthes hamata* and 4th year arable crop (sorghum on heavier soils; pearl millet on lighter soils) Silvi-pastoral system as above
	Medium(0-45 cm)	Ley farming or Horti-pastoral system	Mango, *Sapota,* Guava, *Aonla, Stylosanthes hamata/Macroptilium atropurpureum*

Selection of top-feed tree species and understorey crops

1. Tree components

The tree should have the following essential features, *viz.*,

1. Fast growing, having multipurpose uses, eg. fodder, fuel, timber *etc.*

2. High palatability and digestability of foliage.

3. Good coppicing ability.

4. Ability to withstand browsing, trampling and intensive lopping.

5. Resistant/tolerant to drought and extremes of temperature.

2. Pasture component

To be compatible with tree species, the pasture component should fulfil the following requirements.

1. It should be able to grow well and be compatible with other forage crops.
2. It should be tolerant to drought and extremes of temperature.
3. Must be prolific seeder and in case of non seeding types, it should be able to propagate vegetatively.
4. Possess high palatability with good nutritive value.
5. Should have the ability to conserve soil, water and nutrients.
6. It should be able to withstand over-grazing and trampling.

List of suitable grasses and legumes for drought prone areas

Grasses	Legumes
Cenchrus ciliaris (**Buffel grass**)	*Stylosanthes hamata* (**Stylo**)
Cenchrus setigerus (**Bird wood grass**)	*Stylosanthes scabra* (**Stylo**)
Dicanthium annulatum	*Dolichos lab lab* (**Field Bean**)
Bothrichloa insculpa	*Macroptilum atropurpureum* (**Siratro**)
Chloris gayana (**Rhodes grass**)	*Cyamopsis tetragonoloba* (**Guar**)
Sorghum sudanense (**Sudan grass**)	*Vigna unguiculata* (**Cowpea**)
Panicum antidotale	*Dolichos biflorus* (**Horsegram**)
Panicum coloralum	*Pennisetum pedicellatum* (**Dinanat grass**)
Panicum maximum (**Guinea grass**)	*Urochloa mosambicensis* (**Sabi grass**)

List of top-feed tree species

Common name	Botanical name
Babool	*Acacia nilotica*
Gum Arabic	*Acacia Senegal*
Maharukh	*Ailanthus excelsa*
Khairwal	*Bauhinea purpurea*
Mopane	*Colophospermum mopane*
Shisham	*Dalbergia sissoo*
Gumhar	*Gmelina arboren*
Dhaman	*Grewia optiva*
Anjan	*Hardwickia binata*
Subabul	*Leucaena leucocephala*

Khejri	*Prosopis cineraria*
Vilayati kikar	*Parkinsonia aculeate*
Jangal jalebi	*Pithecellobium dulce*
Safeda, Fras	*Populus* sp.
Agati	*Sesbania grandflora*
Farash	*Tamarix aphylla*

Agri-horticultural systems

Due to continued misuse of land, its productivity has considerably gone down. Lands have been exposed to severe erosion, thereby reducing the inflow of precipitation into soil and consequently resulting in poor ground water recharge. However, due to pressure of population, degraded marginal lands are being cropped. Agroforestry is a set of land use technologies that would be useful in restoring the productivity of such sites. Tree farming helps in controlling erosion, increasing water storage and increasing biological productivity on degraded sites. With this premise, coupled with dependence of rural poor people on these lands, the potential for agroforestry system in rehabilitation of degraded lands appear promising. Agro-horticultural system is one form of agroforestry where the tree component is a fruit tree.

Advantages of agri-horticultural systems

This system mainly focuses on higher income per unit area. The economic aspect of fruit culture is also no less important. Well maintained and established orchards bring better returns than field crops from the same piece of land. The question arises, then, as to why the development of fruit industry has been so slow in spite of the large varieties of fruits that are being grown. The answer lies in fruit growing as a long-term venture. The enterprise requires a high initial investment and a high recurring expenditure. Most of the fruit trees are woody perennials with deep roots that take a minimum number of years to bear fruit. They require intensive cultivation, including specialized methods of propagation, irrigation and plant protection. The horticultural products are perishable and need careful transport, handling and marketing. Although the expenses involved are more than that of raising an annual crop, the farmer can be compensated to a great extent by intercropping during the early stages and later by judicious management of trees.

This system also helps to generate more employment for the farmer especially during the off-season when the crops are not cultivated. The management of the tree is important to get the best returns from not only the tree but also the under-crops. Fruit growing also provides scope for ancillary industries like fruit processing, canning, preservation, dehydration, essential oils, package, transport and refrigeration.

The other advantages of the system arise from the fruit trees mainly. It is very well recognized now that fruits should not be considered a luxury, since they are protective foods necessary for the maintenance of human health. Many fruits are a rich source of energy giving carbohydrates. In addition, fruits are more valuable source of minerals, vitamins and enzymes which are required for better functioning of human body and for fighting diseases. Besides these, pectin and cellulose found in several fruits stimulate the intestinal activity and protect the human body against various diseases. Many fruits are also known to possess specific medicinal values. This is more important for a country like ours where the majority of the population is vegetarian.

9

Tillage, Tilth, Frequency and Depth of Cultivation – Compaction in Soil Tillage, Concept of Conservation Tillage

Tillage

The mechanical manipulation of soil with tools and implements for obtaining conditions ideal for seed germination, seedling establishment and growth of crops is known as tillage. Tillage may be described as the practice of modifying the state of the soil in order to provide conditions favourable to crop growth.

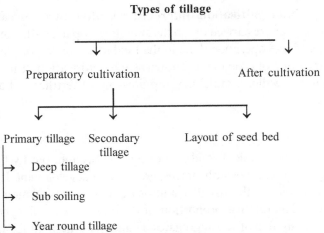

1) Preparatory tillage: Tillage operations that are carried out from the time of harvest of a crop to the sowing of the next crop are known as preparatory cultivation/ tillage. *OR* Operations carried out in any cultivated land to prepare seedbed for sowing crops are preparatory tillage. It includes primary and secondary tillage.

a) Primary tillage: It mainly includes the ploughing operation which is opening of the compacted soil with the help of different ploughs. Ploughing is done to:

1) Open the hard soil,

2) Separate the top soil from lower layers,

3) Invert the soil whenever necessary and

4) Uproot the weeds and stubbles.

b. Secondary tillage : Lighter or finer operation performed on the soil after primary tillage are known as secondary tillage which includes the operations performed after ploughing, leveling, discing, harrowing etc.

2. Seedbed preparation: When the soil is brought to a condition suitable for germination of seeds and growth of crops, called as seedbed.

After preparatory tillage the land is to be laid out properly for irrigating crops if irrigation is available for sowing or planting seeding which are known as seedbed preparation: It includes harrowing, leveling, compacting the soil, preparing irrigation layouts such as basins, borders, rides and furrows etc. and carried out by using hand tools or implements like harrow, rollers plank, rider etc. After field preparation, sowing is done with seed drills. Seeds are covered and planking is done so as to level and impart necessary compaction.

3. Inter tillage/ inter cultivation/ intercultural/ after care operation: The tillage operations that are carried out in the standing crop are called inter tillage operations. The tillage operation done in the field after sowing or planting and prior to the harvesting of crop plants known as inter cultivation. It includes gap filling , thinning , weeding , mulching, top dressing of fertilizers, hoeing, and earthing up etc.

Tilth

Tilth is defined as the physical condition of the soil brought out by tillage that influences crop emergence, establishment, growth and development. It indicates two properties of soil *viz.* the size distribution of aggregates and mellowness or friability of soil. The relative proportion of different sized soil aggregates is known as size distribution of soil aggregates. Higher percent of larger aggregates (>5 mm in diameter) is necessary for irrigated agriculture, while higher percent of smaller aggregates (1-2 mm in diameter) is desirable for dry land agriculture. The size distribution of aggregates depends on soil type, soil moisture content (at which ploughing is done) and subsequent cultivation. Mellowness or friability is that property of soil by which the clods when dry become more crumbly.

Objectives of tillage in drylands are:

1. Develop desired soil structure for a seed bed which allows rapid infiltration and good retention of rainfall.

2. Minimize soil erosion by following practices such as contour tillage, tillage across the slope *etc.*

3. Control weeds and remove unwanted crop plants.

4. Manage crop residues.

5. Obtain specific land configurations for in- situ moisture conservation, drainage, planting *etc.*

6. Incorporate and mix manures, fertilizers, pesticides or soil amendments into the soil.

7. Accomplish segregation by moving soil from one layer to another removal of rocks or root harvesting.

Attention must be paid to the depth of tillage, time of tillage, direction of tillage and intensity of tillage.

1. Depth of tillage

It depends on soil type, crop and time of tillage

a) Deep tillage: of 25-30 cm is beneficial for deep heavy clay soils to improve permeability and to close cracks formed with hard pans, deep tillage once in 2-3 years with chisel plough up to 35-45 cm depth at 60-120 cm.

b) Medium deep tillage: of 15-20 cm depth is generally sufficient for most soils and crops. It is recommended for medium deep soils, shallow rooted crops, soils with pan free horizon and for stubble incorporation.

c) Shallow tillage: up to 10 cm is followed in light textured soils, and shallow soils and in soils highly susceptible to erosion. In soils prone for surface crusting, shallow surface stirring or shallow harrowing is useful.

2. Time of tillage

Early completion of tillage is often helpful to enable sowing immediately after rainfall and before the soil dries up. Summer tillage or off-season tillage done with pre season rainfall causes more conservation of moisture and also enables early and timely sowing. It is particularly useful for pre-monsoon sowing.

3. Direction of tillage

For moisture conservation, ploughing across the slope or along the contour is very effective. Plough furrows check the velocity of runoff, promote more infiltration and improve soil moisture storage.

4. Intensity of tillage

It refers to the number of times tillage is done. Frequent ploughing in shallow light textured soils will pulverize the soils into fine dust and increase the susceptibility to erosion. In heavy soils, leaving the land in a rough and cloddy stage prior to sowing is useful for more depression storage.

Modern concepts of tillage

In dry lands, rainfall is received simultaneously over a large area. In order to ensure timely sowing before soil dries up, the interval between land preparations and sowing must be narrowed down. This calls for completion of tillage over a large area in quick time. Dependence on bullock power and traditional wooden plough may not help in this regard. Use of more efficient tillage implements and mechanization of tillage operations are warranted.

Minimum/optimum/reduced tillage: It is the tillage system aimed at reducing the number of tillage operations to the minimum level i.e. necessary for better seed bed preparation, rapid germination for maintenance of optimum plant stand. It not only saves time, energy and cost but also helps in moisture conservation. The objectives of these systems include (1) reducing energy input and labour requirement for crop production (2) conserving soil moisture and reducing erosion (3) providing optimum seedbed rather than homogenizing the entire soil surface, and (4) keeping field compaction to minimum. The advantages are:

 i) Reduction of soil compaction

 ii) Reduction of soil erosion

 iii) Increases infiltration of water

 iv) Increased soil fertility due to decomposition of crop residues

 v) Less cost of production because less number of tillage operations.

Forms of minimum tillage

 i) Row zone tillage.

 ii) Plough plant tillage.

 iii) Wheel track planting.

b) Conservation/mulch tillage.

c) Zero tillage or no-till system.

1. Zero tillage

It is an extreme form of minimum tillage where primary tillage is completely avoided and secondary tillage is restricted to crop zone. In this method use of machinery and herbicides with relatively low or no residual effect on the crop to be established will play a major role. The machinery should have attachments for four operations namely, cleaning the narrow strip over crop row, open the soil for seed insertion, placing the seed and covering the seed.

Advantages are:

i) Increases the biological activity in the soil.

ii) Organic matter content of the soil is increase due to decomposition of crop residues.

iii) Reduction of surface runoff.

Disadvantages are:

i) Poor seed germination.

ii) High dose of N required for mineralization.

iii) Some perennial weeds and voluntary plants predominate.

iv) More disease and pest incidence.

Conservation agriculture

Conservation agriculture (CA) defined as minimal soil disturbance (no-till, NT) and permanent soil cover (mulch) combined with rotations, is a recent agricultural management system that is gaining popularity in many parts of the world. Cultivation is defined by the Oxford English dictionary as 'the tilling of land', 'the raising of a crop by tillage' or 'to loosen or break up soil'. Other terms used in this dictionary include 'improvement or increase in soil fertility'. All these definitions indicate that cultivation is synonymous with tillage or ploughing.

The other important definition that has been debated and defined in many papers is the word 'sustainable'. The Oxford English dictionary defines this term as 'capable of being borne or endured, upheld, defended, maintainable'. Something that is sustained is 'kept up without intermission or flagging, maintained over a long period'. This is an important concept in today's agriculture, since the human race will not want to compromise the ability of its future offspring to produce

182 Rainfed Agriculture

their food needs by damaging the natural resources used to feed the population today.

2. Cultivation techniques or tillage

The history of tillage dates back many millennia when humans changed from hunting and gathering to more sedentary and settled agriculture mostly in the Tigris, Euphrates, Nile, Yangste and Indus river valleys. Reference to ploughing or tillage is found from 3000 BC in Mesopotamia. Historical development of agriculture with tillage being a major component of management practices. With the advent of the industrial revolution in the nineteenth century, mechanical power and tractors became available to undertake tillage operations; today, an array of equipment is available for tillage and agricultural production. The following summarizes the reasons for using tillage.

1. Tillage was used to soften the soil and prepare a seedbed that allowed seed to be placed easily at a suitable depth into moist soil using seed drills or manual equipment. This results in good uniform seed germination.

2. Wherever crops grow, weeds also grow and compete for light, water and nutrients. Every gram of resource used by the weed is one less gram for the crop. By tilling their fields, farmers were able to shift the advantage from the weed to the crop and allow the crop to grow without competition early in its growth cycle with resulting higher yield.

3. Tillage helped release soil nutrients needed for crop growth through mineralization and oxidation after exposure of soil organic matter to air.

4. Previous crop residues were incorporated along with any soil amendments (fertilizers, organic or inorganic) into the soil. Crop residues, especially loose residues, create problems for seeding equipment by raking and clogging.

5. Many soil amendments and their nutrients are more available to roots if they are incorporated into the soil; some nitrogenous fertilizers are also lost to the atmosphere if not incorporated.

6. Tillage gave temporary relief from compaction using implements that could shatter below-ground compaction layers formed in the soil.

7. Tillage was determined to be a critical management practice for controlling soilborne diseases and some insects.

There is no doubt that this list of tillage benefits was beneficial to the farmer, but at a cost to him and the environment, and the natural resource base on which farming depended. The utility of ploughing was first questioned by a forward looking agronomist in the 1930s, Edward H. Faulkner, in a manuscript called 'Ploughman's Folly'. In a foreword to a book entitled 'Ploughman's folly and a second look' by EH Faulkner, Paul Sears notes that: Faulkner's genius was to question the very basis of agriculture itself—the plough. He began to see that the curved mouldboard of the modern plough, rather than allowing organic matter to be worked into the soil by worms and other burrowing animals, instead buries this valuable material under the subsoil where it remains like a wad of undigested food from a heavy meal in the human stomach. The tragic dust storm in the mid-western United States in the 1930s was a wake-up call to how human interventions in soil management and ploughing led to unsustainable agricultural systems. In the 1930s, it was estimated that 91 Mha of land was degraded by severe soil erosion; this area has been dramatically reduced today.

Conservation tillage and conservation agriculture

Since the 1930s, during the following 75 years, members of the farming community have been advocating a move to reduced tillage systems that use less fossil fuel, reduce run-off and erosion of soils and reverse the loss of soil organic matter. The first 50 years was the start of the conservation tillage (CT) movement and, today, a large percentage of agricultural land is cropped using these principles. However, in the book 'No-tillage seeding', explained 'as soon as the modern concept of reduced tillage was recognized, everyone, it seems, invented a new name to describe the process'. The book goes on to list 14 different names for reduced tillage along with rationales for using these names.

In the book CT is defined as: the collective umbrella term commonly given to no-tillage, direct-drilling, minimum-tillage and/or ridge-tillage, to denote that the specific practice has a conservation goal of some nature. Usually, the retention of 30% surface cover by residues characterizes the lower limit of classification for conservation-tillage, but other conservation objectives for the practice include conservation of time, fuel, earthworms, soil water, soil structure and nutrients. Thus residue levels alone do not adequately describe all conservation tillage practices.

This has led to confusion among the agricultural scientists and, more importantly, the farming community. To add to the confusion, the term 'Conservation Agriculture' has recently been introduced by the Food and Agriculture Organization, and others, and its goals defined by FAO as follows:

Conservation agriculture (CA) aims to conserve, improve and make more efficient use of natural resources through integrated management of available soil, water and biological resources combined with external inputs. It contributes to environmental conservation as well as to enhanced and sustained agricultural

184 Rainfed Agriculture

production. It can also be referred to as resource efficient or resource effective agriculture.

This obviously encompasses the 'sustainable agricultural production' need that all mankind obviously wishes to achieve. But this term is often not distinguished from CT. The FAO mentions in its CA website that: Conservation tillage is a set of practices that leave crop residues on the surface which increases water infiltration and reduces erosion. It is a practice used in conventional agriculture to reduce the effects of tillage on soil erosion. However, it still depends on tillage as the structure forming element in the soil. Never the less, conservation tillage practices such as zero tillage practices can be transition steps towards conservation agriculture.

Conservation agriculture defined

The FAO has characterized CA as follows: Conservation agriculture maintains a permanent or semi-permanent organic soil cover. This can be a growing crop or dead mulch. Its function is to protect the soil physically from sun, rain and wind and to feed soil biota. The soil micro-organisms and soil fauna take over the tillage function and soil nutrient balancing. Mechanical tillage disturbs this process. Therefore, zero or minimum tillage and direct seeding are important elements of CA. A varied crop rotation is also important to avoid disease and pest problems.

The three key principles of CA are:

1. Permanent residue soil cover,
2. Minimal soil disturbance and
3. Crop rotations.

The principles of conservation agriculture

Conservation agriculture emphasizes that the soil is a living body, essential to sustain quality of life on the planet. In particular, it recognizes the importance of the upper 0-20 cm of soil as the most active zone, but also the zone most vulnerable to erosion and degradation. Most environmental functions and services that are essential to support terrestrial life on the planet are concentrated in the micro, meso, and macro fauna and flora which live and interact in this zone. It is also the zone where human activities of land management have the most immediate, and potentially the greatest impact. By protecting this critical zone, we ensure the health, vitality, and sustainability of life on this planet.

The principles of CA and the activities to be supported are described as follows:

1. Maintaining permanent soil cover and promoting minimal mechanical disturbance of soil through zero tillage systems, to ensure sufficient living and/or residual biomass to enhance soil and water conservation and control soil erosion. In turn, this improves soil aggregation, soil biological activity and soil biodiversity, water quality, and increases soil carbon sequestration. Also, it enhances water infiltration, improves soil water use efficiency, and provides increased insurance against drought. Permanent soil cover is maintained during crop growth phases as well as during fallow periods, using cover crops and maintaining residues on the surface.

2. Promoting a healthy, living soil through crop rotations, cover crops, and the use of integrated pest management technologies. These practices reduce requirements for pesticides and herbicides, control off-site pollution, and enhance biodiversity. The objective is to complement natural soil biodiversity and to create a healthy soil micro environment that is naturally aerated, better able to receive, hold and supply plant available water, provides enhanced nutrient cycling, and better able to decompose and mitigate pollutants. Crop rotations and associations can be in the form of crop sequences, relay cropping, and mixed crops.

3. Promoting application of fertilizers, pesticides, herbicides, and fungicides in balance with crop requirements. Feed the soil rather than fertilize the crop. This will reduce chemical pollution, improve water quality, and maintain the natural ecological integrityof the soil, while optimizing crop productivity and economic returns.

4. Promoting precision placement of inputs to reduce costs, optimize efficiency of operations, and prevent environmental damage. Treat problems at the field location where they occur, rather than blanket treatment of the field, as with conventional systems. Benefits are increased economic and field operation efficiencies, improved environmental protection, and reduced (optimized) input costs. Precision is exercised at many levels:seed, fertilizer and spray placement; permanent wheel placement to stop random compaction; individual weed killing with spot-spraying rather than field spraying, etc. Global positioning systems are sometimes used to enhance precision, but farmer sensibility in problem diagnosis and precise placement of treatments is the principal basis. In small-scale farming systems and horticultural systems, it also includes differential plantings on hills and ridges to optimize soil moisture and sunshine conditions;

5. Promoting legume fallows (including herbaceous and tree fallows where suitable), composting and the use of manures and other organic soil amendments. This improves soil structure and biodiversity, and reduces the need for inorganic fertilizers.

186   Rainfed Agriculture

6. Promoting agroforestry for fiber, fruit and medicinal purposes. Agroforestry (trees on farms) provides many opportunities for value added production, particularly in tropical regions, but these technologies are also used as living contour hedges for erosion control, to conserve and enhance biodiversity, and to promote soil carbon sequestration. Conservation agriculture strives to develop a balanced coexistence between rural and urban societies, based on increased urban awareness of the environmental benefits and services provided by the rural sector. It works with the international and national market place to develop financial mechanisms to ensure that environmental benefits provided by CA are recognized by society at large, and benefits accrued to CA practitioners. A recent example is the marketing of carbon credits under the Kyoto Accord, but this is only the beginning. Many other opportunities for environmental payments will develop in the future, including the potential for farm products produced under a new "conservation label". The rapid adoption of conservation technologies by large as well as small farmers in many areas of the world, often without government support, is clear evidence of the economic, environmental and social benefits that accrue from these practices.

Seeding practices in dryland area

1. Establishment of optimum population: Poor or suboptimal population is a major reason for low yields in rainfed crops. Establishment of an optimum population depends on-

    a) Seed treatment and seed hardening

    b) Sowing at optimum soil moisture

    c) Time of sowing

    d) Depth of sowing

    e) Method of sowing

    f) Crop geometry

a) Seed treatment

Seed treatment is done for many purposes such as protection against pests and diseases, inoculation of bio-fertilizers and inducing drought tolerance.

Seed hardening

It is done to induce drought tolerance in emerging seedlings. It is the process of soaking seeds in chemical solution and drying to induce tolerance to drought.

Soil moisture stress immediately after sowing affects germination and establishment.

b) Sowing at optimum soil moisture

An effective rainfall of 20-25 mm which can wet a depth of 10-15 cm is needed for sowing. Moisture stress at or immediately after sowing adversely affects germination and establishment of seedlings. To ensure adequate soil moisture at sowing, sowing has to be done as early as possible after soaking rainfall is received.

c) Time of sowing

Optimum time of sowing is indicated by adequate rainfall to wet seeding depth and continuity of rainfall after sowing. The probable sowing time in a rainfed area is the week which has a rainfall of not less than 20 mm with coefficient of variability less than 100% and the probability of a wet week following wet week. Timely sowing ensures optimal yield besides it may also help pest avoidance. In Maharashtra *kharif* sorghum cultivated in 30 lakh hectares and more than 70% under hybrid prone to shoot fly. If sown at early July, the pest incidence can be avoided.

Pre-monsoon dry seeding

In some regions, where heavy clay soils dominate, sowing after rains impossible due to high stickiness of soil. Here sowing is done in dry soil, 2-3 weeks before the onset of monsoon (pre-monsoon). Seeds will remain in soil and germinate only on receipt of optimum rainfall.

The advantages of pre-monsoon dry seeding are

 i) Early sowing

 ii) Uniform germination and good establishment

 iii) Utilization of first rainfall itself for germination instead of for land preparation in post monsoon sowing

 iv) Early maturity before closure of monsoon and avoidance of stress at maturity.

The success of pre-monsoon dry seeding depends on the following

 i) It is recommended for bold seeds like cotton and sorghum only and not for all crops.

 ii) Time of advance sowing must be fixed based on rainfall analysis for date of onset of monsoon and continuity of rainfall after sowing.

iii) Seeds must be hardened to ensure quick germination and drought tolerance.

iv) Seeding depth must be such that seeds will germinate only after receipt of rainfall to wet that depth is received. Surface sowing may lead to germination with less rainfall and death due to subsequent soil drying.

v) Off season tillage is necessary to enable sowing in dry soil before monsoon.

vi) Seed damage by soil insects has to be prevented.

Examples of pre-monsoon sowing

1. For sorghum in black soils, pre-monsoon dry seeding is recommended 1-2 weeks before onset of monsoon with depth of sowing at 5 cm and seed hardening with 2 per cent potassium di-hydrogen phosphate or potassium chloride.

2. For cotton in black soils, pre-monsoon dry seeding is recommended at 2-4 weeks before commencement of monsoon, with a sowing depth of 5 cm and seed hardening with CCC (500 ppm) or potassium chloride or DAP at 2% level.

d) Optimum depth of sowing

When seeds are sown on surface or at very shallow depth, germination and seedling growth are affected when surface soil moisture dries up. Sowing at a depth where soil moisture availability is adequate, ensure early and uniform germination and seedling establishment. Optimum depth of sowing varies with crop, especially seed size and penetration power of plumule. Sesame 1-2 cm; pearl millet and minor millets 2-3 cm; pulses, sorghum, sunflower 3-5 cm; cotton, maize 5 cm; coriander 7 cm.

e) Method of sowing

In dry lands, it is important to sow the seeds in moist soil layer to ensure proper germination and seedling emergence. Dibbling of seeds and planting of seedlings are also adopted for some crops (cotton, tobacco, chilies). The choice of sowing method depends on seed size, soil condition, time available, cropping system, crop geometry, sowing depth, source of power, cost of sowing, etc.

f) Crop geometry

Crop geometry refers to the shape of land occupied by individual plants as decided by spacing between rows and between plants. It depends on the root spread and the canopy size of the crop and the cropping system.

Setline cultivation

It is a form of minimum tillage practice predominant in Saurastra region of India where farmers are adopting the practice of continuously cultivating, manuring and sowing wide spaced crops in the same line or row year after year. In between the rows the soil is worked with blade or harrowed only for weed control. The crops like sorghum, bajra, cotton, and groundnut are cultivated by this method.

The advantages are

i) Reduced cost of cultivation.

ii) As the crops are raised in the same row, the rhizosphere is loose with good aeration and permeability without development of hard pans in the sub soil.

Soil crusts

Soil crusts are hard layers that develop at the soil surface due to action of rain drop or irrigation water and subsequent drying. Soil crusts often hinder the emergence of seedlings and hence establishment of crop.

Depth and intensity of ploughing

The purpose of cultivation is to control weeds and conserve soil moisture. Deep ploughing gives better response in fields infested with weeds. This is also practiced to incorporate the residues, particularly in sandy soils. Incorporation of residues in the deep layer having fine texture may improve the chemical properties of the soil. Under dryland conditions, deep ploughing improves soil moisture content. However, yield advantages due to deep ploughing depend on rainfall and type of crop. Normally, yield advantage is reported during normal and above normal rainfall years. Depth of ploughing mainly depends on the effective root zone depth of the crops. Generally, crops with tap root system require greater depth of ploughing, while fibrous, shallow-rooted crops require shallow ploughing.

It is important to practice deep ploughing for long duration, deep-rooted crops. Root crops generally respond better to deep tillage than shallow-rooted crops like cereals. Potato generally responds better to deep tillage, particularly in clayey and loamy soils. Sugarbeet also shows positive response to deep ploughing. Deep tillage is also practised for crops like sugarcane. Small grain crops such as millets, sesame and flax do as well with shallow as with deep ploughing. Deep cultivation is important to break the compact layers. Compaction occurs due to use of tractors in wet soil condition. Chiseling up to 75 cm helps to break the sub-soil compaction.

Deep ploughing helps to break the soil compaction and increase water absorption and root penetration, resulting in improved plant growth. The number of ploughings necessary to obtain a good tilth depends on soil type, weed problem and crop residues on the soil surface. In heavy soils, more number of ploughings is necessary, the range being 3 to 5 ploughings. Light soils require 1 to 3 ploughings to obtain proper tilth of the soil. When weed growth and plant residues are higher, more number of ploughings is necessary

Tillage practices in dryland areas

Tillage for soil conservation: Tillage is an important and primary tool for conservation of the land. As per definition, its primary purpose is to provide a favorable soil environment for the plant growth which is indirectly related to the soil conservation. The effect of tillage on soil erosion is the function of its several effects on soil such as aggregation surface sealing infiltration and resistant to erosion destruction of soil structure either by excessive tillage or tillage operations at improper soil moisture condition tends to increase the soil erodibility, causing significant soil loss.

To achieve a best result for soil conservation the following points should be considered for tillage operations.

1. Till no more than necessary.

2. Till only when soil moisture is in the favorable limit; and

3. Vary the depth of ploughing.

Types of soil conservation tillage practices

Mulch tillage

1. The mulch intercepts the falling raindrops over the land surface and thus dissipating their kinetic energy which result in reduction or elimination of their dispersing action on the soil structure.

2. The mulch tillage increases the infiltrate capacity.

3. The obstacles caused by leaves, stems and roots over the field reduce the velocity of surface flow and thus controlling the sheet erosion.

4. It maintains the soil relatively cool and moist which are essential for good plant growth.

5. Increases the crop yield by developing several conducive effects on soil.

Mulching materials

The followings are used as mulching materials

- Cut grasses or foliage
- Straw materials
- Wood chips
- Saw dusts
- Papers
- Sand stones
- Glass woods
- Metal foils
- Stones
- Plastics.

Types of mulches

The mulches may be of following types

a) Natural

b) Synthetic

c) Petroleum

d) Conventional

e) Inorganic and

f) Organic.

Soil conservation – agronomic measures

Soil conservation is using and managing the land based on the capabilities of the land itself involving application of the best management practices leading to profitable crop production without land degradation.

Control of water erosion

Water erosion occurs simultaneously in two steps: detachment of soil particles by falling raindrops and transportation of detached particles by flowing water. Hence preventing the detachment of soil particles and their transportation can minimize water erosion. Principles of water erosion control are

- Maintenance of soil infiltration capacity
- Soil protection from rainfall
- Control of surface runoff and
- Safe disposal of surface runoff.

For a sound soil conservation programme every piece of land must be used in accordance with the land capability classification.

Measures of water erosion control

1. Agronomic measures
2. Mechanical measures (Engineering measures)
3. Forestry measures
4. Agrostological measures.

Agronomic measures

In soil and water conservation programmes agronomic measures have to be considered in co-ordination with others for their effectiveness. These measures are effective in low rainfall areas particularly in fairly erosion resistant soils having gentle slope (< 2 %). The different agronomic measures include:

1. Land preparation
2. Contour cultivation
3. Choice of crops
4. Strip cropping
5. Crop rotation / cropping systems
6. Cover crops
7. Mulching
8. Application of manures and fertilizers
9. Application of chemicals.

1) Land preparation: Land preparation including post harvest tillage influence intake of water, obstruction to surface flow and consequently the rate of erosion. Deep ploughing or chiseling has been found effective in reducing erosion. Rough cloddy surface is also effective in reducing erosion.

2) Contour cultivation (contour farming): A line joining the points of equal elevation is called contour. All the cultural practices such as ploughing, sowing, intercultivation etc. done across the slope reduce soil and water loss. By ploughing and sowing across the slope, each ridge of plough furrow and each row of the crop act as obstruction to the runoff and provide more time for water to enter into the soil leading to reduced soil and water loss.

3) Choice of crops: Row crops or tall growing crops such as sorghum, maize, pearl millet etc., are not effective in conserving soil as they expose majority of the soil and hence they are known as erosion permitting crops. Whereas close growing crops such as cowpea, groundnut, green gram, black gram etc., which protect soil are known as erosion resisting crops as they are very effective in reducing soil loss by minimizing the impact of rain drop and acting as obstruction to runoff.

4) Strip cropping: It is a system of growing of few rows of erosion resisting crops and erosion permitting crops in alternate strips on contour (across the slope) with the objective of breaking long slopes to prevent soil loss and runoff. Close growing erosion resisting crops reduce the transporting and eroding power of water by obstructing runoff and filtering sediment from runoff to retain in the field. The width of the erosion permitting and erosion resisting crops vary as per the slope of the field. The strip cropping resembles the intercropping. With increase in per cent slope of the soil, the width of erosion permits and erosion resisting crops decreases. The normal ratio between the erosion resisting crops and erosion permitting crops is 1: 3. Among the different crops the anti- erosion value of pillipesara is highest, where as cotton crop recorded the lowest value.

i) Contour strip cropping: The erosion permitting crops and erosion resisting crops are grown in alternate strips along the contours.

ii) Field strip cropping: Alternate strips of erosion permitting crops and erosion resisting crops are raised across the general slope not necessarily on exact contour

iii) Wind strip cropping: Strip cropping of erosion permitting and erosion resisting crops across the direction of the most prevailing wind irrespective of the contour.

iv) Buffer strip cropping: This type of strip cropping is practiced in areas having steep slopes and badly eroded soils where strips of permanent cover crops or perennial legumes or grasses or shrubs are alternated with field crops.

The strip cropping is simple, cheap and effective soil conservation practice and can be adopted by the farmers.

5) Crop rotation / cropping system: Monocropping of erosion permitting crops accelerates soil and water loss year after year. Intercropping of erosion permitting crops and erosion resisting crops or their rotation has been found effective for reducing soil and water loss. Inclusion of legumes like lucerne in crop rotation reduces soil loss even in soils having 13% slope.

6) Cover crops: Good ground cover by canopy gives the protection to the land like an umbrella and minimize soil erosion. Besides conserving soil and moisture, the cover crops hold those soluble nutrients, which are lost by leaching. The third advantage of the cover crops is the addition of organic matter. The legumes provide better cover and better protection. Among the legumes cowpea has been found to produce maximum canopy followed by horsegram, green gram, black gram and dhaincha.

7) Mulching: Mulching of soil with available plant residues reduce soil loss considerably by protecting the soil from direct impact of raindrop and reducing the sediment carried with runoff. A minimum plant residue cover of 30 per cent is necessary to keep runoff and soil loss within the acceptable limits. Vertical mulching also reduces soil loss particularly in vertisols by increasing infiltration.

8) Application of manures and fertilizers: Organic manures besides supplying nutrients improve soil physical conditions thereby reduce soil loss. Fertilizers improve vegetative canopy, which aid in erosion control.

9) Use of chemicals: Breakdown of aggregates by the falling raindrops is the main cause of detachment of soil particles. Soils with stable aggregates resist breakdown and thus resist erosion. Aggregate stability can be increased by spraying chemicals like polyvinyl alcohol @ 480 kg/ha (rate will depend on the type of soil). Soils treated with bitumen increase water stable aggregates and infiltration capacity of the soil.

Techniques and practices of soil moisture conservation

In dryland agriculture, soil moisture is the most limiting factor. It is lost as evaporation from the soil surface and as transpiration from the plant surfaces. The evapotranspiration losses can be reduced by:

1. Mulches
2. Antitranspirants

Tillage, Tilth, Frequency and Depth of Cultivation 195

3. Wind breaks

4. Weed control

Mulching

About 60-75% of rainfall is lost through evaporation. These evaporation losses can be reduced by applying mulches. Mulch is any material applied on the soil surface to check evaporation and improve soil water. Application of mulches results in additional benefits like soil conservation, moderation of temperature, reduction in soil salinity, weed control and improvement of soil structure.

Effect of mulches on soil properties

- *Soil water*: Mulches improve soil water by reduction of evaporation, runoff and weeds and increase infiltration. Application of mulches on the soil surface obstructs the solar radiation reaching the soil. It also checks the escape of water vapour by physical obstruction. Mulch slows flow velocity of runoff.

- *Soil temperature*: The effects on soil temperature are highly variable and depends on the type of mulch material. White or reflective type of plastic mulches decrease soil temperature or have no effect. Transparent plastic mulches increase soil temperature.

- *Soil salinity*: Due to mulch infilteration increases and evaporation decreases, the salts do not accumulate in the surface layers.

Types of mulches

- *Soil mulch or dust mulch*: If the soil is loosened, it acts as a mulch for reducing evaporation. This loose surface soil is called soil mulch or dust mulch. Its usefulness is doubtful in alfisols but helps in closing deep cracks in vertisols.

- *Stubble mulch*: Crop residues like wheat straw or cotton stalks etc., are left on soil surface as a stubble mulch. The advantages of stubble mulch farming are protection of soil from erosion and reduction of evaporation losses.

- *Straw mulch*: If straw is used as mulch, it is called straw mulch.

- *Plastic mulch*: Plastic materials like polyethylene, polyvinyl chloride are also used as mulching materials.

- *Vertical mulching*: It consists of digging narrow trenches across the slope at intervals and placing the straw or crop residues in these trenches. This is mostly practiced in coffee gardens. It prolongs the beneficial effect of subsoiling.

10

Soil Erosion: Definition, Nature and Extent of Erosion, Types of Erosion and Factors Affecting Soil Erosion

Definition

Soil erosion is the process of detachment of soil particles from the top soil and transportation of the detached soil particles by wind and / or water. The agents causing erosion are wind and water. The detaching agents are falling raindrop, channel flow and wind. The transporting agents are flowing water, rain splash and wind.

Erosion is the process by which soil and rock are removed from the Earth's surface by exogenic processes such as wind or water flow, and then transported and deposited in other locations.

While erosion is a natural process, human activities have increased by 10-40 times the rate at which erosion is occurring globally. Excessive erosion causes problems such as desertification, decreases in agricultural productivity due to land degradation, sedimentation of waterways, and ecological collapse due to loss of the nutrient rich upper soil layers. Water and wind erosion are now the two primary causes of land degradation; combined, they are responsible for 84% of degraded acreage, making excessive erosion one of the most significant global environmental problems.

Industrial agriculture, deforestation, roads, anthropogenic climate change and urban sprawl are amongst the most significant human activities in regard to their effect on stimulating erosion. However, there are many prevention and remediation practices that can curtail or limit erosion of denuded soils.

Nature and extent of erosion

The problem of soil erosion exists all over the country. Out of the 329 m ha of India's geographical area about 175 m ha (53.3%) is subjected to soil erosion and some kind of land degradation. About 150 m ha is subjected to wind and water erosion. It is estimated that about 5333 Mt of soil is detached annually. Out of this 29% is carried away by rivers to seas and about 10% is deposited in reservoirs resulting in 1-2% of loss of storage capacity annually. The estimated annual soil loss is 16.35 tones/ha/year.

Physiographically India is divided into three regions as follows:

a) Himalayan region: Geologic immaturity made this region more vulnerable to erosion. High degree of seismicity of the area, very steep slopes, weak geological formation and improper land use practices accelerate erosion losses. Gullying, land slides and slips are most common.

b) Gangetic plains: Major problems in the region are riverine erosion, drainage, saline and alkali soil conditions.

c) Peninsular region: Main problems of this region are rill and gully erosion. Arid regions have severe wind erosion. Semi-arid regions are subjected to sheet and gully erosion and ravines are serious problem in Yamuna and Chambal region. Floods and stream bank cutting and sand deposition have degraded lands of north east region with heavy rainfall. South and south east are characterized by undulating terrain with severe erosion in black and red laterite soils.

Losses due to erosion

i)   Loss of fertile top soil

ii) Loss of rain water

iii) Loss of nutrients

iv) Silting up of reservoirs

v)   Damage to forests

vi) Reduction in soil depth

vii) Floods

viii) Adverse effect on public health

ix) Loss of fertile land

x)   Economic losses

Types of erosion

There are two major types of soil erosion

a) Geological erosion (Natural or normal erosion): is said to be in equilibrium with soil forming process. It takes place under natural vegetative cover completely undisturbed by biotic factors. This is very slow process.

b) Accelerated erosion: is due to disturbance in natural equilibrium by the activities of man and animals through land mismanagement, destructing of forests over grazing etc. Soil loss through erosion is more than the soil formed due to soil forming process.

Based on the agents causing erosion, erosion is divided into a. Water erosion b. Wind erosion c. Wave erosion

Water erosion

Loss of soil from land surface by water including runoff from melted snow and ice is usually referred to as water erosion. Major erosive agents in water erosion are impacting/ falling raindrops and runoff water flowing over soil surface.

Process of water erosion

Detachment of soil particles is by either raindrop impact or flowing water. Individual raindrops strike the soil surface at velocities up to 9 m/s creating very intensive hydrodynamic force at the point of impact leading to soil particle detachment. Over land flow detaches soil particles when their erosive hydrodynamic force exceeds the resistance of soil to erosion. Detached soil particles are transported by raindrop splash and runoff. The amount of soil transported by runoff is more than due to raindrop splash. Thus the falling raindrops break the soil aggregates and detach soil particles from each other. The finer particles (silt and clay) block the soil pores and increase the rate of runoff and hence loss of water and soil.

Forms of water erosion

Water erosion occurs in stages identified as sheet erosion, rills, gullies, ravines, landslides and stream bank erosion.

a) Sheet erosion: It is the uniform removal of surface soil in thin layers by rainfall and runoff water. The breaking action of raindrop combined with surface flow is the major cause of sheet erosion. It is the first stage of erosion and is least conspicuous, but most extensive.

b) Rill erosion: When runoff starts, channelisation begins and erosion is no longer uniform. Raindrop impact does not directly detach any particles below flow line in rills but increases the detachment and transportation capacity of the flow. Rill erosion starts when the runoff exceeds 0.3 to 0.7 mm/s. Incisions are formed on the ground due to runoff and erosion is more apparent than sheet erosion. This is the second stage of erosion. Rills are small channels, which can be removed by timely normal tillage operations.

c) Gully erosion: It is the advanced stage of water erosion. Size of the unchecked rills increase due to runoff. Gullies are formed when channelised runoff form vast sloping land is sufficient in volume and velocity to cut deep and wide channels. Gullies are the spectacular symptoms of erosion. If unchecked in time no scope for arable crop production.

d) Ravines: They are the manifestations of a prolonged process of gully erosion. They are typically found in deep alluvial soils. They are deep and wide gullies indicating advanced stage of gully erosion.

e) Landslides: Landslides occur in mountain slopes, when the slope exceeds 20% and width is 6m. Generally land slides cause blockage of traffic in ghat roads.

f) Stream bank erosion: Small streams, rivulets, torrents (hill streams) are subjected to stream bank erosion due to obstruction of their flow. Vegetation sprouts when streams dry up and obstructs the flow causing cutting of bank or changing of flow course.

Factors affecting water erosion

a) Climate: Water erosion is directly a function of rainfall and runoff. Amount, duration and distribution of rainfall influences runoff and erosion. High intensity rains of longer duration causes severe erosion. Greater the intensity, larger the size of the raindrop. Rainfall intensity more than 5 cm/hr is considered as severe. Total energy of raindrops falling over a hectare land with rainfall intensity of 5 cm /hr is equal to 625 H.P. This energy can lift 89 times the surface 17.5 cm of soil from one ha to a height of 3 ft. Two- thirds of the above energy is used for sealing soil pores. Runoff may occur without erosion but there is no water erosion without runoff. The raindrop thus breaks down soil aggregates, detaches soil particles and leads the rainwater with the fine particles. These fine particles seal the pores of the surface soil and increases runoff causing erosion.

b) Topography: The degree, length and curvature of slope determine the amount of runoff and extent of erosion. Water flows slowly over a gentle slope where as at a faster rate over a steeper one. As water flows down the slope, it accelerates under the forces of gravity. When runoff attains a velocity of about 1 m/s it is capable of eroding the soil. If the percent of slope is increased by 4 times the velocity of water flowing down is doubled. Doubling the velocity quadruples the erosive power and increases the quantity of soil that can be transported by about 32 times and size of the particles that can be transported by about 64 times.

c) Vegetation: Vegetation intercepts the rainfall and reduces the impact of raindrops. It also decreases the velocity of runoff by obstructing the flow of water. The fibrous roots are also effective in forming stable soil aggregates, which increases infiltration and reduces erosion.

d) Soil properties: Soil properties that influence soil erodability by water may be grouped into two types.

i. Those properties that influence the infiltration rate and permeability

ii. Those properties that resist the dispersion, splashing, abrasion and transporting forces of rainfall and runoff.

The structure, texture, organic matter and moisture content of upper layers determine the extent of erosion. Sandy soils are readily detachable but not readily transportable. Soils of medium to high clay content have low infiltration capacities and they are readily transported by water after they are dispersed, but their detachability is generally low.

Man and beast

Man and beast accelerates erosion by extensive farming and excessive grazing. Faulty practices like cultivation on steep slopes, cultivation up and down the slope, felling and burning of forests etc., leads to heavy erosion. Excessive grazing destroys all vegetation and increases the erosion.

Estimation of soil loss by water erosion

Based on the mechanism and factors influencing soil erosion, a universal soil loss equation (USLE) developed by Wischmeier (1959) is most useful for predicting soil loss due to water erosion. It is an empirical equation and estimates average annual soil loss per unit area as a function of major factors affecting sheet and rill erosion. It enables determination of land management erosion rate relationships for a wide range of rainfall, soil slope and crop and management conditions and to select alternative cropping and management combinations that limit erosion rates to acceptable limits.

$$A = R \times K \times L \times S \times C \times P$$

where, A= predicted soil loss in t/ha/year

R= rainfall erosivity factor or index

K= soil erodibility factor

L= length of slope factor

S= slope steepness factor

C= soil cover and management factor and

P= erosion control factor

Wind erosion

Erosion of soil by the action of wind is known as wind erosion. It is a serious problem on lands devoid of vegetation. It is more common in arid and semi-arid regions. It is essentially a dry weather phenomenon stimulated by the soil moisture deficiency. The process of wind erosion consists of three phases: initiation of movement, transportation and deposition. About 33 m ha in India is affected by wind erosion. This includes 23.49 m ha of desert and about 6.5 m ha of coastal sands. The Thar Desert is formed mainly by blow in sand.

Mechanism of wind erosion

Lifting and abrasive action of wind results in detachment of tiny soil particles from the granules or clods. The impact of these rapidly moving particles dislodge other particles from clods and aggregates. These dislodged particles are ready for movement. Movement of soil particles in wind erosion is initiated when the pressure by the wind against the surface soil grains overcomes the force of gravity on the grains. Minimum wind velocity necessary for initiating the movement of most erodable soil particles (about 0.1 mm diameter) is about 16 km /hr at a height of 30.5 cm. Most practical limit under field conditions, where a mixture of sizes of single grained material present is about 21 km/hr at a height of 30.5 cm. In general movement of soil particles by wind takes place in three stages: saltation, surface creep and suspension.

a. Saltation: It is the first stage of movement of soil particles in a short series of bounces or jumps along the ground surface. After being rolled by the wind, soil particles suddenly leap almost vertically to form the initial stage of movement in saltation. The size of soil particles moved by saltation is between 0.1 to 0.5 mm in diameter. This process may account for 50 to 70% of the total movement by wind erosion.

b. Surface creep: Rolling and sliding of soil particles along the ground surface due to impact of particles descending and hitting during saltation is called surface creep. Movement of particles by surface creep causes an abrasive action of soil surface leading to break down of non-erodable soil aggregates. Coarse particles longer than 0.5 to 2.0 mm diameter are moved by surface creep. This process may account for 5 to 25% of the total movement.

c. Suspension: Movement of fine dust particles smaller than 0.1 mm diameter by floating in the air is known as suspension. Soil particles carried in suspension are deposited when the sedimentation force is greater than the force holding the particles in suspension. This occurs with decrease in wind velocity. Suspension usually may not account for more than 15% of total movement.

Physical processes affecting erosion

Rainfall

A spoil tip covered in rills and gullies due to erosion processes caused by rainfall. There are four primary types of erosion that occur as a direct result of rainfall: splash erosion, sheet erosion, rill erosion, and gully erosion. Splash erosion is generally seen as the first and least severe stage in the soil erosion process, which is followed by sheet erosion, then rill erosion and finally gully erosion (the most severe of the four).

In splash erosion, the impact of a falling raindrop creates a small crater in the soil, ejecting soil particles. The distance these soil particles travel can be as much as two feet (0.6 m) vertically and five feet (1.5 m) horizontally on level ground. Once the rate of rainfall is faster than the rate of infiltration into the soil, surface runoff occurs and carries the loosened soil particles down the slope.

Sheet erosion is the transport of loosened soil particles by overland flow.

Rill erosion refers to the development of small, ephemeral concentrated flow paths which function as both sediment source and sediment delivery systems for erosion on hillslopes. Generally, where water erosion rates on disturbed upland areas are greatest, rills are active. Flow depths in rills are typically of the order of a few centimeters or less (around an inch) and slopes may be quite steep. This means that rills exhibit hydraulic physics very different from water flowing through the deeper, wider channels of streams and rivers.

Gully erosion occurs when runoff water accumulates and rapidly flows in narrow channels during or immediately after heavy rains or melting snow, removing soil to a considerable depth.

Rivers and streams

Valley erosion is occurring due to the flow of the stream, and the boulders and stones (and much of the soil) that are lying on the edges are glacial till that was left behind as ice age glaciers flowed over the terrain.

Valley or stream erosion occurs with continued water flow along a linear feature. The erosion is both downward, deepening the valley, and headward, extending the valley into the hillside, creating head cuts and steep banks. In the earliest stage of stream erosion, the erosive activity is dominantly vertical, the valleys have a typical 'V' cross-section and the stream gradient is relatively steep. When some base level is reached, the erosive activity switches to lateral erosion, which widens the valley floor and creates a narrow flood plain. The stream gradient becomes nearly flat, and lateral deposition of sediments becomes important as the stream meanders across the valley floor. In all stages of stream erosion, by far the most erosion occurs during times of flood, when more and faster-moving water is available to carry a larger sediment load. In such processes, it is not the water alone that erodes: suspended abrasive particles, pebbles and boulders can also act erosively as they traverse a surface, in a process known as traction.

Bank erosion is the wearing away of the banks of a stream or river. This is distinguished from changes on the bed of the watercourse, which is referred to as scour. Erosion and changes in the form of river banks may be measured by inserting metal rods into the bank and marking the position of the bank surface along the rods at different times.

Thermal erosion is the result of melting and weakening permafrost due to moving water. It can occur both along rivers and at the coast. Rapid river channel migration observed in the Lena River of Siberia is due to thermal erosion, as these portions of the banks are composed of permafrost-cemented non-cohesive materials. Much of this erosion occurs as the weakened banks fail in large slumps. Thermal erosion also affects the Arctic coast, where wave action and near-shore temperatures combine to undercut permafrost bluffs along the shoreline and cause them to fail.

Coastal erosion

Shoreline erosion, which occurs on both exposed and sheltered coasts, primarily occurs through the action of currents and waves but sea level (tidal) change can also play a role.

Hydraulic action takes place when air in a joint is suddenly compressed by a wave closing the entrance of the joint. This then cracks it. *Wave pounding* is when the sheer energy of the wave hitting the cliff or rock breaks pieces off. *Abrasion* or *corrasion* is caused by waves launching seaload at the cliff. It is the most effective and rapid form of shoreline erosion (not to be confused with *corrosion*). *Corrosion* is the dissolving of rock by carbonic acid in sea water. Limestone cliffs are particularly vulnerable to this kind of erosion. *Attrition* is where particles/seaload carried by the waves are worn down as they hit each other and the cliffs. This then makes the material easier to wash away. The material ends up as shingle and sand. Another significant source of erosion, particularly on carbonate coastlines, is the boring, scraping and grinding of organisms, a process termed *bioerosion*.

Sediment is transported along the coast in the direction of the prevailing current (long shore drift). When the upcurrent amount of sediment is less than the amount being carried away, erosion occurs. When the upcurrent amount of sediment is greater, sand or gravel banks will tend to form as a result of deposition. These banks may slowly migrate along the coast in the direction of the longshore drift, alternately protecting and exposing parts of the coastline. Where there is a bend in the coastline, quite often a build up of eroded material occurs forming a long narrow bank (a spit). Armoured beaches and submerged offshore sandbanks may also protect parts of a coastline from erosion. Over the years, as the shoals gradually shift, the erosion may be redirected to attack different parts of the shore.

Glaciers

Glaciers erode predominantly by three different processes: abrasion/scouring, plucking, and ice thrusting. In an abrasion process, debris in the basal ice scrapes along the bed, polishing and gouging the underlying rocks, similar to sandpaper on wood. Glaciers can also cause pieces of bedrock to crack off in the process of plucking. In ice thrusting, the glacier freezes to its bed, then as it surges forward, it moves large sheets of frozen sediment at the base along with the glacier. The erosion caused by glaciers worldwide has been shown to erode mountains so effectively that the term glacial buzz-saw has become widely used, which describes the limiting effect of glaciers on the height of mountain ranges. As mountains grow higher, they generally allow for more glacial activity (especially above the glacial equilibrium line altitude), which causes increased rates of erosion of the mountain, decreasing mass faster than isostatic rebound can add to the mountain. This provides a good example of a negative feedback loop. Ongoing research is showing that while glaciers tend to decrease mountain size, in some areas, glaciers can actually reduce the rate of erosion, acting as a glacial armor.

206   Rainfed Agriculture

These processes, combined with erosion and transport by the water network beneath the glacier, leave moraines, drumlins, ground moraine (till), kames, kame deltas, moulins, and glacial erratics in their wake, typically at the terminus or during glacier retreat.

Floods

At extremely high flows, kolks, or vortices are formed by large volumes of rapidly rushing water.

Freezing and thawing

Cold weather causes water trapped in tiny rock cracks to freeze and expand, breaking the rock into several pieces. This can lead to gravity erosion on steep slopes. The scree which forms at the bottom of a steep mountainside is mostly formed from pieces of rock (soil) broken away by this means. It is a common engineering problem wherever rock cliffs are alongside roads, because morning thaws can drop hazardous rock pieces onto the road.

Mass movement

Mass movement is the downward and outward movement of rock and sediments on a sloped surface, mainly due to the force of gravity. Mass movement is an important part of the erosional process, and is often the first stage in the breakdown and transport of weathered materials in mountainous areas. It moves material from higher elevations to lower elevations where other eroding agents such as streams and glaciers can then pick up the material and move it to even lower elevations. Mass-movement processes are always occurring continuously on all slopes; some mass-movement processes act very slowly; others occur very suddenly, often with disastrous results.

Exfoliation

Exfoliation is a type of erosion that occurs when a rock is rapidly heated up by the sun. This results in the expansion of the rock. When the temperature decreases again, the rock contracts, causing pieces of the rock to break off. Exfoliation occurs mainly in deserts due to the high temperatures during the day and cold temperatures at night.

Lightning strikes

When water in cracked rock is rapidly heated by a lightning strike, the resulting steam explosion can erode rock and shift boulders. It may be a significant factor in erosion of tropical and subtropical mountains that have never been glaciated. Evidence of lightning strikes includes craters, partially melted rock and erratic magnetic fields.

Factors affecting erosion rates

Precipitation and wind speed

Climatic factors include the amount and intensity of precipitation, the average temperature, as well as the typical temperature range, seasonality, wind speed, and storm frequency. In general, given similar vegetation and ecosystems, areas with high-intensity precipitation, more frequent rainfall, more wind, or more storms are expected to have more erosion.

Rainfall intensity is the primary determinant of erosivity, with higher intensity rainfall generally resulting in more erosion. The size and velocity of rain drops is also an important factor. Larger and higher-velocity rain drops have greater kinetic energy, and thus their impact will displace soil particles by larger distances than smaller, slower-moving rain drops.

Soil structure and composition

The composition, moisture, and compaction of soil are all major factors in determining the erosivity of rainfall. Sediments containing more clay tend to be more resistant to erosion than those with sand or silt, because the clay helps bind soil particles together. Soil containing high levels of organic materials are often more resistant to erosion, because the organic materials coagulate soil colloids and create a stronger, more stable soil structure. The amount of water present in the soil before the precipitation also plays an important role, because it sets limits on the amount of water that can be absorbed by the soil (and hence prevented from flowing on the surface as erosive runoff). Wet, saturated soils will not be able to absorb as much rain water, leading to higher levels of surface runoff and thus higher erosivity for a given volume of rainfall. Soil compaction also affects the permeability of the soil to water, and hence the amount of water that flows away as runoff. More compacted soils will have a larger amount of surface runoff than less compacted soils.

Vegetative cover

Vegetation acts as an interface between the atmosphere and the soil. It increases the permeability of the soil to rainwater, thus decreasing runoff. It shelters the soil from winds, which results in decreased wind erosion, as well as advantageous changes in microclimate. The roots of the plants bind the soil together, and interweave with other roots, forming a more solid mass that is less susceptible

208 Rainfed Agriculture

to both water and wind erosion. The removal of vegetation increases the rate of surface erosion.

Topography

The topography of the land determines the velocity at which surface runoff will flow, which in turn determines the erosivity of the runoff. Longer, steeper slopes (especially those without adequate vegetative cover) are more susceptible to very high rates of erosion during heavy rains than shorter, less steep slopes. Steeper terrain is also more prone to mudslides, landslides, and other forms of gravitational erosion processes.

Human activities that increase erosion rates

Agricultural practices

Unsustainable agricultural practices are the single greatest contributor to the global increase in erosion rates. The tillage of agricultural lands, which breaks up soil into finer particles, is one of the primary factors. The problem has been exacerbated in modern times, due to mechanized agricultural equipment that allows for deep, which severely increases the amount of soil that is available for transport by water erosion. Others include mono-cropping, farming on steep slopes, pesticide and chemical fertilizer usage (which kill organisms that bind soil together), row-cropping, and the use of surface irrigation. A complex overall situation with respect to defining nutrient losses from soils, could arise as a result of the size selective nature of soil erosion events. Loss of total phosphorus, for instance, in the finer eroded fraction is greater relative to the whole soil. Extrapolating this evidence to predict subsequent behaviour within receiving aquatic systems, the reason is that this more easily transported material may support a lower solution 'P' concentration compared to coarser sized fractions. Tillage also increases wind erosion rates, by dehydrating the soil and breaking it up into smaller particles that can be picked up by the wind. Exacerbating this is the fact that most of the trees are generally removed from agricultural fields, allowing winds to have long, open runs to travel over at higher speeds. Heavy grazing reduces vegetative cover and causes severe soil compaction, both of which increase erosion rates.

Deforestation

In this clearcut, almost all of the vegetation has been stripped from surface of steep slopes, in an area with very heavy rains. Severe erosion occurs in cases such as this, causing stream sedimentation and the loss of nutrient rich topsoil.

In an undisturbed forest, the mineral soil is protected by a layer of leaf litter and an humus that cover the forest floor. These two layers form a protective mat over the soil that absorbs the impact of rain drops. They are porous and highly

Soil Erosion: Definition, Nature and Extent of Erosion, Types of Erosion 209

permeable to rainfall, and allow rainwater to slow percolate into the soil below, instead of flowing over the surface as runoff. The roots of the trees and plants hold together soil particles, preventing them from being washed away. The vegetative cover acts to reduce the velocity of the raindrops that strike the foliage and stems before hitting the ground, reducing their kinetic energy. However it is the forest floor, more than the canopy, that prevents surface erosion. The terminal velocity of rain drops is reached in about 8 metres (26 feet). Because forest canopies are usually higher than this, rain drops can often regain terminal velocity even after striking the canopy. However, the intact forest floor, with its layers of leaf litter and organic matter, is still able to absorb the impact of the rainfall.

Deforestation causes increased erosion rates due to exposure of mineral soil by removing the humus and litter layers from the soil surface, removing the vegetative cover that binds soil together, and causing heavy soil compaction from logging equipment. Once trees have been removed by fire or fogging, infiltration rates become high and erosion low to the degree the forest floor remains intact. Severe fires can lead to significant further erosion if followed by heavy rainfall.

Globally one of the largest contributors to erosive soil loss in the year 2006 is the slash and burn treatment of tropical forests. In a number of regions of the earth, entire sectors of a country have been rendered unproductive. For example, on the Madagascar high central plateau, comprising approximately ten percent of that country's land area, virtually the entire landscape is sterile of vegetation, with gully erosive furrows typically in excess of 50 metres (160 ft) deep and 1 kilometre (0.6 miles) wide. Shifting cultivation is a farming system which sometimes incorporates the slash and burn method in some regions of the world. This degrades the soil and causes the soil to become less and less fertile.

Roads and urbanization

Urbanization has major effects on erosion processes—first by denuding the land of vegetative cover, altering drainage patterns, and compacting the soil during construction; and next by covering the land in an impermeable layer of asphalt or concrete that increases the amount of surface runoff and increases surface wind speeds. Much of the sediment carried in runoff from urban areas (especially roads) is highly contaminated with fuel, oil, and other chemicals. This increased runoff, in addition to eroding and degrading the land that it flows over, also causes major disruption to surrounding watersheds by altering the volume and rate of water that flows through them, and filling them with chemically polluted sedimentation. The increased flow of water through local waterways also causes a large increase in the rate of bank erosion.

Climate change

The warmer atmospheric temperatures observed over the past decades are expected to lead to a more vigorous hydrological cycle, including more extreme rainfall events. The rise in sea levels that has occurred as a result of climate change has also greatly increased coastal erosion rates. Studies on soil erosion suggest that increased rainfall amounts and intensities will lead to greater rates of erosion. Thus, if rainfall amounts and intensities increase in many parts of the world as expected, erosion will also increase, unless amelioration measures are taken. Soil erosion rates are expected to change in response to changes in climate for a variety of reasons. The most direct is the change in the erosive power of rainfall. Other reasons include:

a) Changes in plant canopy caused by shifts in plant biomass production associated with moisture regime;

b) Changes in litter cover on the ground caused by changes in both plant residue decomposition rates driven by temperature and moisture dependent soil microbial activity as well as plant biomass production rates;

c) Changes in soil moisture due to shifting precipitation regimes and evapo-transpiration rates, which changes infiltration and runoff ratios;

d) Soil erodibility changes due to decrease in soil organic matter concentrations in soils that lead to a soil structure that is more susceptible to erosion and increased runoff due to increased soil surface sealing and crusting;

e) A shift of winter precipitation from non-erosive snow to erosive rainfall due to increasing winter temperatures;

f) Melting of permafrost, which induces an erodible soil state from a previously non-erodible one

g) Shifts in land use made necessary to accommodate new climatic regimes.

Studies by Pruski and Nearing indicated that, other factors such as land use not considered, it is reasonable to expect approximately a 1.7% change in soil erosion for each 1% change in total precipitation under climate change.

Global environmental effects

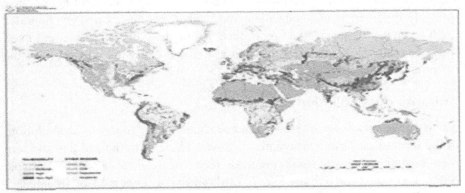

World map indicating areas that are vulnerable to high rates of water erosion.

Due to the severity of its ecological effects, and the scale on which it is occurring, erosion constitutes one of the most significant global environmental problems we face today.

Land degradation

Water and wind erosion are now the two primary causes of land degradation; combined, they are responsible for 84% of degraded acreage. Each year, about 75 billion tons of soil is eroded from the land—a rate that is about 13-40 times as fast as the natural rate of erosion. Approximately 40% of the world's agricultural land is seriously degraded. According to the United Nations, an area of fertile soil the size of Ukraine is lost every year because of drought, deforestation and climate change. In Africa, if current trends of soil degradation continue, the continent might be able to feed just 25% of its population by 2025, according to UNU's Ghana-based Institute for Natural Resources in Africa.

The loss of soil fertility due to erosion is further problematic because the response is often to apply chemical fertilizers, which leads to further water and soil pollution, rather than to allow the land to regenerate.

Sedimentation of aquatic ecosystems

Soil erosion (especially from agricultural activity) is considered to be the leading global cause of diffuse water pollution, due to the effects of the excess sediments flowing into the world's waterways. The sediments themselves act as pollutants, as well as being carriers for other pollutants, such as attached pesticide molecules or heavy metals. The effect of increased sediments loads on aquatic ecosystems can be catastrophic. Silt can smother the spawning beds of fish, by filling in the space between gravel on the stream bed. It also reduces their food supply, and causes major respiratory issues for them as sediment enters their gills. The biodiversity of aquatic plant and algal life is reduced, and invertebrates are also unable to survive and reproduce. While the sedimentation event itself might be relatively short-lived, the ecological disruption caused by the mass die off often persists long into the future.

Airborne dust pollution

Soil particles picked up during wind erosion are a major source of air pollution, in the form of airborne particulates—"dust". These airborne soil particles are often contaminated with toxic chemicals such as pesticides or petroleum fuels, posing ecological and public health hazards when they later land, or are inhaled/ingested. Dust from erosion acts to suppress rainfall and changes the sky color from blue to white, which leads to an increase in red sunsets. Dust events have been linked to a decline in the health of coral reefs.

Tectonic effects

The removal by erosion of large amounts of rock from a particular region, and its deposition elsewhere, can result in a lightening of the load on the lower crust and mantle. This can cause tectonic or isostatic uplift in the region.

Monitoring, measuring, and modeling erosion

Monitoring and modeling of erosion processes can help us better understand the causes, make predictions, and plan how to implement preventative and restorative strategies. However, the complexity of erosion processes and the number of areas that must be studied to understand and model them (*e.g.* climatology, hydrology, geology, chemistry, physics, *etc.*) makes accurate modelling quite challenging. Erosion models are also non-linear, which makes them difficult to work with numerically, and makes it difficult or impossible to scale up to making predictions about large areas from data collected by sampling smaller plots.

The most commonly used model for predicting soil loss from water erosion is the Universal Soil Loss Equation (USLE), which estimates the average annual soil loss as :

$$A = RKLSCP$$

where R is the rainfall erosivity factor, K is the soil erodibility factor, L and S are topographic factors representing length and slope, and C and P are cropping management factors. A new soil erosion model named G2 monitors soil erosion by a spatio-temporal index. G2 is a dynamic model, as it takes account of contemporary changes of rainfall erosivity and vegetation retention. Based on the empirical USLE-family models, it needs calibration for rainstorm erosivity, while vegetation retention is based on biophysical parameters derived with remote sensing. Erosion is measured and further understood using tools such as the micro-erosion meter (MEM) and the traversing micro-erosion meter (TMEM). The MEM has proved helpful in measuring bedrock erosion in various ecosystems around the world. It can measure both terrestrial and oceanic erosion. On the other hand, the TMEM can be used to track the expanding and contracting of volatile rock formations and can give a reading of how quickly a rock formation is deteriorating.

Prevention and remediation

The most effective known method for erosion prevention is to increase vegetative cover on the land, which helps prevent both wind and water erosion. Terracing is an extremely effective means of erosion control, which has been practiced for thousands of years by people all over the world. Windbreaks (also called shelterbelts) are rows of trees and shrubs that are planted along the edges of agricultural fields, to shield the fields against winds. In addition to significantly reducing wind erosion, windbreaks provide many other benefits such as improved microclimates for crops (which are sheltered from the dehydrating and otherwise damaging effects of wind), habitat for beneficial bird species, carbon sequestration, and aesthetic improvements to the agricultural landscape. Traditional planting methods, such as mixed-cropping (instead of monocropping) and crop rotation have also been shown to significantly reduce erosion rates.

11

Drainage Considerations and Agronomic Management; Rehabilitation of Abandoned Jhum Lands and Measures to Prevent Soil Erosion

---

Drainage - definition

Agricultural drainage is the artificial removal and safe disposal of excess water either from the land surface or soil profile, more specifically, the removal and safe disposal of excess gravitational water from the crop root zone to create favourable conditions for crop growth to enhance agricultural production.

Benefits of drainage

a) It provides better soil environment for plant growth by creating favourable soil aeration conditions.

b) It improves the soil structure and in turn increases the soil infiltration .

c) High infiltration capacity reduces soil erosion.

d) It hastens the warming of the soils and maintains desirable soil temperature, which accelerates plant growth and bacterial activity.

e) It promotes increased leaching of salts and prevents accumulation of salts in the crop root zone.

f) In well drained soils, less time and less labour are required for tillage operations.

Problems or effects of ill-drainage

a) Limitation of aeration.

b) Accumulation of CO_2 and toxic substances like H_2S, ferrous sulfide *etc* in the crop root zones.

c) Reduced water uptake due to reduced activity of roots as a result of oxygen stress.

d) Reduced nutrient uptake.

e) Development of soil salinity and alkalinity.

f) Anaerobic condition and prevalence of plant diseases.

g) Stunted plant growth and development which results in reduced yield.

Types of drainage

Broadly drainage systems are of two types- Surface and Sub-surface.

1. Surface drainage systems

Safe removal and disposal of excess water primarily from land surface or cropped area by a net work of surface drains or constructed channels and through proper land shaping is known as surface drainage. There are four general types of surface drainage systems used in flat areas having a slope of <2% viz.,

(a) Random drain system (b) Parallel field drain system (c) Parallel open ditch system and (d) Bedding system

a) Random drain system

This system is usually adopted in areas where the ground surface is characterized by a series and depression (undulating land surface) and where small depressions are to be drained off (Fig. 1). Depending upon the possibility the field drains are designed in such a way to connect one depression to another and water is safely conveyed to lateral drains. These lateral drains ultimately guide the water to main outlet drain. The field drains besides occupying the land area are likely to interfere with farm operations.

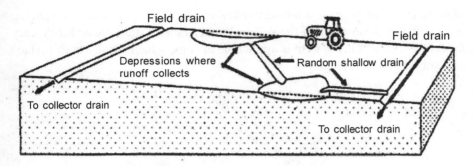

Fig.1 : Random field drain system

b) Parallel field drain system

The parallel field ditch system is used in places where the surface is uniform and has few noticeable ridges or depressions (Fig. 2). In this system the surface of individual fields is graded in such a way so that the runoff water drains into field drains, which in turn discharge water into field laterals bordering the field and finally the laterals in turn lead water into the main outlet ditch through protected over falls. Laterals and mains should be deeper than field drains to provide free out-fall. Maximum spacing of parallel field drains is about 200 m for sandy soils and about 100 m for clay soils. It is the most desirable surface drainage method and is well suited both for irrigated and rainfed areas.

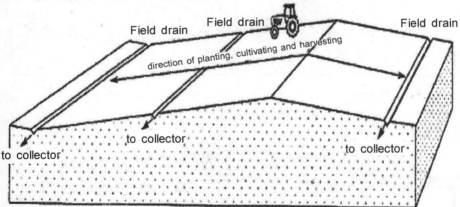

Fig. 2 : Parallel field drain system

c) Parallel open ditch system

The parallel open ditch system is similar to parallel field drain system in all respects except that the drains are replaced by open ditches which are comparatively deeper and have steeper side slopes than the field drains (Fig. 3).

Maximum length of grade draining to ditch should not be > 180 m. The spacing of the ditches depends upon the soil and water table conditions and may vary from 60 – 200m. This system is applicable in soils, which require both surface and sub-surface drainage.

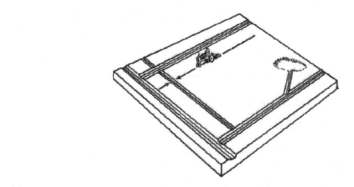

Fig. 3. : Parallel open ditch system

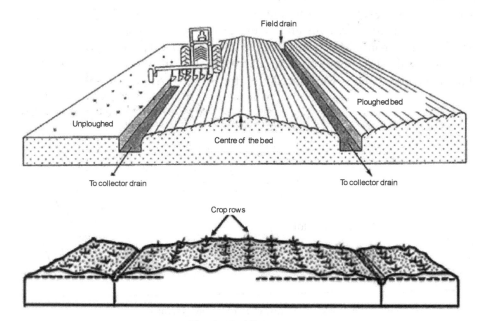

Fig. 4 : Bedding system

d) Bedding system

This system is usually adopted in fields with very little slope, usually 0.5% or less and slowly permeable soils. It is essentially a tillage operation wherein the land is ploughed into a series of parallel beds separated by dead furrows, which run in the direction of greatest slope lateral drains are located perpendicular to slope (Fig. 4). The ploughing operations are to be carried out parallel to the furrows. The bed width and length varies between 8 to 30 m and 10 to 300m respectively depending upon field conditions i.e., land use, slope, soil permeability and farming operations. While bed height should not exceed 40 cm.

2. Sub-surface drainage systems

The removal and safe disposal of excess water that has already entered the soil profile is considered sub-surface drainage. Though several sub-surface systems are available, the most commonly used and effective ones are Tile drainage and Mole drainage systems.

i. Tile drainage systems

Tile drains removes excess water from the soil through a continuous line of tiles (pipes) laid at specified depth and grade. The pipes are made of either concrete or burnt clay. Free water enters through the tile joints and flows out by gravity, so that the water table is lowered below the root zone of the plants.

The common tile drainage system layout followed is: Random or natural system, Parallel lines system and Cut off or intercepting system.

a) Random system

The random system is used in areas that have scattered wet areas somewhat isolated from each other (Fig. 5). Tile lines are laid more or less at random to drain the wet patches.

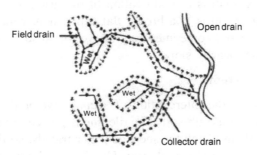

Fig. 5 : Random tile drain system

b) Herringbone system

The system is applicable in places where the main or sub-main is located in a narrow depression *i.e.*, in areas that have a concave surface or a narrow depression with the land sloping to it from both directions (Fig. 6). The parallel laterals enter the sub-main from both sides. It is less economical, because considerable double drainage occurs where the laterals and mains join.

c) Gridiron and parallel systems

The gridiron and parallel systems are similar to that of herringbone system except that the laterals enter the main or sub-main from only one side (Fig. 6). It is the most economical arrangement than herringbone system because one main or sub-main serves as many laterals as possible.

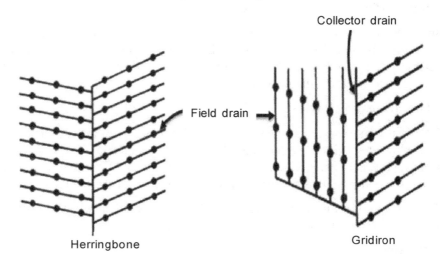

Fig. 6 : Herringbone and Gridiron tile drain systems

d) Double main system

The double main system is a modification of the gridiron system. It may be used where the sub-main is in a broad, flat depression, which frequently is a natural watercourse and sometimes may be wet because of small amounts of seepage water from nearby slopes.

e) Intercepting system

This system involves the interception of seepage water that flows over the surface of an impervious sub-soil. The tile line is placed approximately at the impervious layer along which the seepage water travels, so that water will be intercepted and wet condition is relieved. The tile line should be located in such a way that there is at least 60 cm of soil cover over the top of the tile.

ii. Mole drainage system

Mole drainage is a semi-permanent method of sub-surface drainage, similar to tile drain in layout and operation (Fig. 7). Instead of permanent tiles a continuous circular mole drain (channel) is prepared below the ground surface in the soil profile at desired depth and spacing using a special implement known as mole plough. The depth of the mole drain varies from 4.5 cm to 120 cm depending on the moling equipment and water table.

Diameter of the mole varies from 7.5 to 15 cm. The life the mole drain is 10-15 years. It is adapted to a particular type of soil because the soil stability is more important in this type of sub-surface drainage.

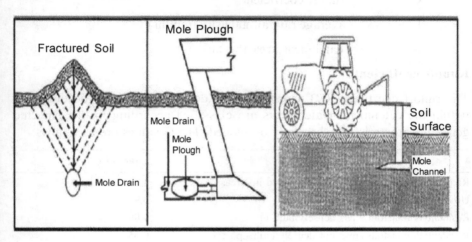

Fig. 7 : Mole plough

Drainage coefficient

It is defined as the depth of water (cm) to be removed in 24 hours period from the entire drainage area. It ranges from 0.6 to 2.5 cm/day and in extreme cases 10 cm/day.

Drainage design and considerations

    a. Computation of peak runoff

    b. Computation of discharge capacity

    c. Design considerations

a. Computation of peak runoff

Rational formula

The rational formula shall be used to compute the peak runoff:

$$Q_r = \frac{1}{360} \, CIA$$

where

Q_r	= peak runoff at the point of design (m³/s)
C	= runoff coefficient
I	= average rainfall intensity (mm/hr)
A	= catchment area (hectares)

Runoff coefficient

The runoff coefficient (C) depends on the degree and type of development within the catchment. Catchments are classified according to the expected general characteristics when fully developed. The C values are as follows:-

	Value of C
Roads, highways, airport runways, paved up areas	1.00
Urban areas fully and closely built up	0.90
Residential/industrial areas densely built up	0.80
Residential/industrial areas not densely built up	0.65
Rural areas with fish ponds and vegetable gardens	0.45

Note : For catchments with composite land use or surface characteristics, a weighted value of C may be adopted.

Rainfall intensity

For a storm of return period (T) years, the rainfall intensity (I) is the average rate of rainfall from such a storm having a duration equal to the time of concentration (tc). The rainfall intensity (I) can be obtained from the Intensity-Duration-Frequency (IDF) curves by estimating the duration of rainfall (equals to the time of concentration, tc) and selecting the required return period of (T) years. The return periods (T) adopted for the design of drainage systems in Singapore are as follows:-

Area Served by Drainage System	Return Period (T)
Catchment of less than 100 ha	10 years
Catchment of 100 to 1000 ha	25 years
Catchment of more than 1000 ha or critical installations	50 to 100 years

Time of concentration

The peak runoff (Qr) occurs at the point of design when all parts of the catchment receiving steady, uniform rainfall intensity are contributing to the outflow at this point. This condition is met when the duration of rainfall equals the time of concentration (tc). The time of concentration (tc) consists of the overland flow time (to) plus the drain flow time from the most remote drainage inlet to the point of design (td), viz. tc = to + td. The overland flow time (to) varies from 5 minutes to 15 minutes, depending on the overland travel distance, land topography and characteristics. The drain flow time (td) shall be estimated from the hydraulic properties of the drainage channel.

b. Computation of discharge capacity

Steady uniform flow condition

Drains are designed for steady uniform flow conditions and one-dimensional method of analysis is used

Manning's formula

Drains shall be designed to have discharge capacities (Qc) adequate to cope with the estimated peak runoffs (Qr). The size, geometry and the bed gradient of a drain determine its discharge capacity (Qc). With the required discharge capacity (Qc) determined [which must be equal to or larger than the peak runoff (Qr)], the size of the drain is computed from the Manning's Formula:-

where

Q_c = discharge capacity of drain (m^3/s)

n   = roughness coefficient

A   = flow area (m^2)

P   = wetted perimeter (m)

R   = A/P = hydraulic radius (m)

S   = bed gradient

Roughness coefficient

The value of the roughness coefficient (n) depends on the drain's flow surface and is given below : -

224   Rainfed Agriculture

Boundary condition	Roughness coefficient (n)
Unplasticised polyvinyl chloride(UPVC)	0.0125
Concrete	0.0150
Brick	0.0170
Earth	0.0270
Earth with stones and weed	0.0350
Gravel	0.0300

Note: Where there are different flow surfaces within a drain section, equivalent roughness coefficient may be used.

c. Design considerations

Minimum velocity and dry weather flow

The velocity of flow in a drain shall not be lower than 1.0 m/s for self-cleansing action to take place. However, the flow rate during dry weather may fall to a low level where this minimum velocity cannot be achieved. The problem can be solved by introducing a small channel in the drain to confine the dry weather flow to a smaller flow section. The dimensions of such a dry weather flow channel depend on the width of the drain

Maximum velocity

The velocity of flow in a drain shall not be too great to cause excessive scouring or hydraulic jumps. Hence the velocity of flow in a concrete-lined drain shall be limited to a maximum of 3.0 m/s or below the critical velocity, whichever is lower. For an earth stream, the maximum velocity shall be limited to 1.5 m/s. Further limitation of the maximum velocity shall be complied with when specified by the Board.

Sub-critical flow

Drains are designed to carry sub-critical flows. Critical state of flow exists when the Froude Number is equal to one. An open channel flow at or near the critical state shall be avoided as under such a condition the water surface is unstable and wavy. In order to secure greater flow efficiency, channel flow shall be designed so that the Froude Number shall fall within the range from 0.8 decreasing to such minimum value as to achieve a practical flow depth and permissible flow velocity.

Freeboard

Freeboard refers to the depth from the top of the drain (cope/bank) to the top of the water surface in the drain at design flow condition. Sufficient freeboard

Drainage Considerations and Agronomic Management; Rehabilitation 225

shall be provided to prevent waves or fluctuation of the water surface from overflowing the cope/bank. Generally, a depth of freeboard equivalent to 15% of the depth of the drain is required.

Agronomic management for waterlogging

Drains in most waterlogging-susceptible cropped areas pay for themselves within a few years. Where drains can only partially overcome the problem, changes to crop species, varieties and management may be necessary. Management options include: choice of crop, seeding, fertiliser, weed and disease control.

The steps to understanding and managing waterlogging are:

1. Identify the problem sites.

2. Consider drainage as the first option for controlling excess water. This will minimize crop losses and land management problems from waterlogging.

3. In all cases, control weeds and plant early using appropriate varieties, high seeding rates and nitrogen.

4. Adjust nitrogen to growing conditions to maximize fertilizer efficiency.

5. Avoid growing crops that are susceptible to water logging.

Management options include: **choice of crop, seeding, fertiliser, weed and disease control**. Typically, with changes to crop rotations and management, major costs would include cost of buying seed and extra fertilizer, and the costs of weed and insect control.

Grain legumes and canola are generally more susceptible to waterlogging than cereals and faba beans. The crops most tolerant of waterlogging include: faba beans, oats, wheat, barley, canola, lupins, peas and chickpeas

1. Seeding crops early and using long-season varieties help to avoid crop damage from waterlogging. Crop damage is particularly severe if plants are waterlogged between germination and emergence. Plant first those paddocks that are susceptible to waterlogging. However, if waterlogging delays emergence and reduces cereal plant density to fewer than 50 plants/m^2, resow the crop.

Increase sowing rates in areas susceptible to waterlogging to give some insurance against uneven germination, and to reduce the dependence of cereal crops on tillering to produce grain. Waterlogging depresses tillering. High sowing rates will also increase the competitiveness of the crop against weeds, which take advantage of stressed crops.

226   Rainfed Agriculture

Crops tolerate waterlogging better with a good nitrogen status before waterlogging occurs. Applying nitrogen at the end of a waterlogging period can be an advantage if nitrogen was applied at or shortly after seeding has been lost by leaching or denitrification. However, nitrogen cannot usually be applied from vehicles when soils are wet, so consider aerial applications.

2. Weed density affect a crop's ability to recover from waterlogging. Weeds compete for water and the small amount of remaining nitrogen, hence the waterlogged parts of a paddock are often weedy. If herbicide resistance is not a problem, spray the weedy areas with a post-emergent herbicide when the paddock is dry enough to allow access, provided the crop is at an appropriate growth stage. Aerial spraying is an alternative when ground-based sprays cannot be used.

3. Root diseases, particularly take-all of wheat and barley, are often more severe in waterlogged crops because the pathogens tolerate waterlogging and low oxygen levels better than the crops. Eliminating grasses from the preceding crops or pastures will reduce the severity of take-all in both well-drained and waterlogged areas. Leaf diseases are likely to be more severe in waterlogged crops because the crop is already under stress. Spraying may be an option after the site has dried, but only in crops with a high yield potential.

Agricultural drainage water management system (ADMS) for improving water quality crop production

Drainage water management can improve water quality and increase crop production. The purpose of an agricultural drainage water management system (ADMS) is to allow for the adjustment of the water table, minimize drainage during times of the year when drainage requirements are reduced, and provide for adequate drainage when needed most. Management of drainage water can provide environmental benefits by reducing the quantity of nutrient enriched drainage water leaving fields, and can provide production benefits by extending the period of time when soil water is available to plants. Water management structures are installed in strategic locations on a field drainage system that provide points of management for the operator. This fact sheet identifies published materials describing agricultural drainage and provides some key considerations for planning and designing a surface, subsurface, or combination surface and subsurface ADMS.

Where to apply the practice

Agricultural drainage is accomplished by a system of surface ditches, subsurface conduits, or by a combination of surface and subsurface components. Drainage systems collect and convey water from fields. Drainage water management can be applied on drained fields where outflows from the drains can be controlled. Some older systems and many newer systems can be adapted to allow for the management of drainage water. Management of drainage water is most effective on systems with pattern drainage, but some systems with random drains can also achieve benefits. If the existing drainage system needs extensive repair or is otherwise not functioning properly it may be necessary to install a new drainage system. When replacing an older drainage system with a new system, make the older system inoperable or incorporate its operation into the new system to avoid undesirable interactions between the two systems. Even where drainage water management is not a goal for new systems, consider planning for future conversion to a managed system.

The topography should be relatively uniform, and flat to gently sloping within a management unit or zone. Non-uniform water table depths can lead to non-uniform crop growth that complicates management decisions. Slopes of 1% or less are recommended. Water management structures should be placed every 1' to 1.5' change in elevation along the drainage ditch or conduit. As the slope increases, more water management structures are required and economic factors and erosion concerns begin to detract from the benefits of the ADMS. A way to minimize the number of water management structures is to install the drains along the contour. Structures should be located on main lines that serve a number of laterals in order to minimize the total number of structures required.

Some agricultural drainage systems are part of a network involving multiple landowners. In such situations, managing a drain on one field can have an adverse impact on operations of adjoining properties. Even without interconnected drainage systems, the impacts of ADMS should be evaluated with respect to adjoining fields.

Water management structures

Retro-fitting an existing subsurface drainage system involves the installation of water management structures. The management mechanism on the structures may be flash boards, gates, valves, risers, and pipes. Flash board risers allow flexibility in manual management of the drainage water. Flashboard type risers can be full-round pipe risers or half-round pipe risers. The full-round risers are used when the control structure is located within the field, while half-round risers are used at the outlet. Gates or valves can be used to temporally stop the flow through the drain, or may be used in such a way that there is a low water table when the gate is opened, and a raised water table when the gate is closed. Gate or valve structures are sometimes automated.

The riser should be large enough to maintain the cross-sectional area of flow from the drain and the length of the flashboard should compensate for the transition from pipe flow to weir flow. It has been a common practice to make the length of the weir a minimum of 1.3 times the diameter of the drain in rectangular risers and 1.7 times the diameter of the drain in circular risers. It is important to size the riser large enough for the easy removal and placement of

228   Rainfed Agriculture

flash boards. Flashboards made of wood can swell when wet and become difficult to remove. It may be more convenient to remove these boards with a chainsaw and replace them with new flashboards.

Water management structures that are not automated must be easily accessible and clearly visible for safety purposes so they are not damaged during field operations. The drainage conduit should be non-perforated within 20 feet of a control structure. Small amounts of seepage at the control structure are usually not a problem. Providing the materials are resistant to damage from ultraviolet light, plastic risers are acceptable except where there is danger of fire, or where they may be damaged by freezing of water surrounding the riser.

Impacts of drainage water management

Drainage water management can have a significant impact on the transport of nitrogen, phosphorous and sediment to surface waters and on crop production. Lowering the water table increases the amount of water passing through the soil. Nitrates and soluble phosphorous move with the drainage water and are transported to the drainage outlets. A lower water table also reduces the frequency and magnitude of surface runoff, and thereby reduces the erosion potential, sediment transport, and the transport of sediment-adsorbed phosphorus. The aerobic conditions created in drained soils decrease the occurrence of denitrification.

Raising the water table decreases the amount of water passing through the soil, and proportionally decreases the transport of nitrates and soluble phosphorous from the field. Raising the water table during the non-growing season can result in a 30% reduction in the discharge of nitrates, but reductions of 50% or greater have often been accomplished. Raising the water table can also increase the amount of surface runoff, leading to increased erosion, sediment transport and transport of sediment-adsorbed phosphorous. Erosion and sediment transport can be controlled with residue management, buffers, grassed waterways, and other conservation practices. Anaerobic conditions created in saturated soils increases the occurrence of denitrification, further reducing nitrate-nitrogen in the drainage water Lowering the water table improves field trafficability and timeliness of crop management operations such as field preparation, planting, and harvesting, and can extend the growing season by allowing earlier access to the field. With a low water table, ponding is less likely to occur or to be sustained when it does occur. A lower water table results in aerobic soil conditions and an increased depth of the root zone. Partially raising the water table after crops are established can conserve soil moisture and may enable a crop to be more productive in the years where there is an extended dry period during the growing season.

A basic recommended strategy

A drainage system infrastructure that enables the operator to manage the water table provides an opportunity to take advantage of the benefits of both high and low water tables. Deciding when to raise or lower a water table can be a difficult decision, particularly when rainfall is uncertain. As with many other practices, more intensive and careful management creates a potential for achieving greater advantages from the system.

In absence of a detailed analysis, there are some basic strategies that can be employed to greatly improve the functionality and benefits of the system. A high water table in the winter months will decrease the transport of nitrates and soluble phosphorus to surface waters. The water table should be lowered in the spring early enough for the field to be accessible for seedbed preparation, planting, and other field operations. Lowering the water table two weeks before field operations in the spring is generally sufficient.

After planting, the water table can be raised to conserve soil moisture for use by the crop during extended dry periods. Once the crop is established, evapotranspiration will often be sufficient to remove excess water from the root zone. It may be necessary to lower the water table during extended wet periods. Careful attention to drainage water management for water conservation may increase yields, particularly in dry years.

In addition to drainage water management, soil, crop, and nutrient management should be a part of a plan to improve water quality in agricultural areas. Nutrient management practices should follow state and local recommendations and NRCS practice standards. Nutrient management applied in conjunction with drainage management can help maximize the effectiveness of both practices.

Field scale monitoring

Monitoring of water table elevations can be accomplished by observing water depths in the water management structures, but in many instances it is better to monitor the water table by establishing monitoring wells in the field. A monitoring well can consist of a perforated PVC pipe that extends below the water table with a measuring rod attached to a float. Other methods for reading water depths include a chalked tape or a string and bobber. Monitoring wells can be located near the edge of the field, where they are easier to monitor and where they are protected from agricultural equipment, but for more representative readings, they should be located farther within the field. Monitoring wells placed between drain lines are more representative of the entire field than those placed directly over a drain line. When they are located in the field they need to be clearly marked and protected from damage by farming operations and livestock. The top of the riser and the measuring rod should be higher than the anticipated crop height during the monitoring periods. In colder climates the monitoring wells may not function over the winter months due to freezing. It may not be necessary to maintain the monitoring wells once the relationship between the water table elevation in the field and the depth of water in the control structure is understood.

Drainage management

1. Operations and timing
2. Equipment used
3. Potential environmental problems
4. Best management practices

Impacts of poor drainage — stunted, yellow plants

The purpose of agricultural drainage is to remove excess water from the soil in order to enhance crop production. In some soils, the natural drainage processes are sufficient for growth and production of agricultural crops, but in many other soils, artificial drainage is needed for efficient agricultural production. Surface drainage is the removal of water that collects on the land surface. Many fields have low spots or depressions where water ponds. Surface drainage techniques such as land leveling, constructing surface inlets to subsurface drains, and the construction of shallow ditches or waterways can allow the water to leave the field rather than causing prolonged wet areas.

Poorly drained area in crop field will damage yields

Subsurface drainage removes excess water from the soil profile, usually through a network of perforated tubes installed 2 to 4 feet below the soil surface. These tubes are commonly called "tiles" because they were originally made from short lengths of clay pipes known as tiles. Water would seep into the small spaces between the tiles and drains away.

Drain tile outlet to a drainage ditch

The most common type of "tile" is corrugated plastic tubing with small perforations to allow water entry. When the water table in the soil is higher than the tile, water flows into the tubing, either through holes in the plastic tube or through the small cracks between adjacent clay tiles. This lowers the water table to the depth of the tile over the course of several days. Drain tiles allow excess water to leave the field, but once the water table has been lowered to the elevation of the tiles, no more water flows through the tiles. In most years, drain tiles are not flowing between June and October.

Operations and timing

On average, about two-thirds of annual precipitation is used by crops in the eastern Corn Belt. The rest falls at a time when it does not meet crop needs. Monthly precipitation remains fairly constant throughout the year, while evapotranspiration (a combination of evaporation from soil and transpiration from the crop), is much higher from June to September. From January to May, and from October to December, precipitation is greater than evapotranspiration, creating a water surplus. The surplus results in excess water in the crop root zone and the need for drainage. Drainage is primarily a concern in the periods prior to the growing season (January to April) so that crops can be planted at the optimum time.

Equipment used

Trenching machine used by drainage contractors to install subsurface drainage tile (shown in white)

Drainage plow being pulled by farm tractor installing a tileline

Designing and installing a drainage system is a complex process. Every field is unique and usually requires an individual design. Drainage depends on topography, crops that will be grown on the field, and soil type. Every soil type has different properties that affect its drainage. Agronomists and engineers have developed recommendations for drainage depth and spacing in specific soil types based on years of experience and knowledge of soil properties. Drainage contractors use these recommendations to design drainage systems that economically and effectively drain a particular field.

Drainage plows that can be pulled by farm tractors are becoming more popular. But most farmers hire contractors to design and install their tile drainage systems because of the knowledge, skills, and experience needed to install a successful system.

Potential environmental concerns

The major concerns related to drainage are:
1. Loss of wetlands, and
2. Increased loss of nitrate through tile drains.

Wetlands

Much of the Midwest landscape consisted of wetlands before large-scale drainage began in the 19th century. Although enormous public health and economic benefits have resulted from the draining of these wetlands over the last 150 years, there have also been negative impacts on the environment. Wetlands have an important hydrologic function in regulating water flow and maintaining water quality, as well as providing habitat for water-based wildlife. Recognition of their value has changed the way our society thinks about and protects wetlands.

Drainage improvements today are rarely for the purpose of converting existing wetlands to agricultural production. Improved drainage is usually aimed at making existing agricultural land more productive. Some fields have drain tiles that were installed 100 or more years ago, and are broken or plugged. In many fields, only a few of the wettest spots were originally drained, while the entire field would benefit from improved drainage. More tiles are often added to improve drainage efficiency, with the goal of increasing production.

Water quality

Poor drainage

Drainage has both positive and negative effects on water quality. In general, less surface runoff, erosion, and phosphorus is lost from land that has good subsurface drainage than from land without drainage improvements or with only surface drainage.

Nitrate loss can be quite high from drained land. Because nitrate is very soluble, it flows easily through the soil and into tile lines. Nitrate flow from subsurface drains is one of the main sources of nitrate in streams and rivers in the Midwest. Concern about hypoxia, or low oxygen levels, in the Gulf of Mexico has increased concern about this nitrate source. Concentrations of nitrate in tile drains are usually quite high (10-40 mg/l).

Pesticides can also flow into subsurface drains, but usually only in very low concentrations. Pesticides move more easily in flow over the soil surface than through the soil, so the highest concentrations of pesticides in tiles are often in fields that have surface inlets into the drains. In fact, subsurface drainage may actually reduce pesticide loss to rivers and streams because it reduces surface runoff.

Best management practices

Traditionally, the goals of drainage were to:

1. Maximize crop yield and
2. Minimize costs of drainage installation.

Reducing water quality effects of drainage is becoming a third objective in drainage design.

Nitrate loss is the biggest water quality concern related to tile drainage. Several new technologies can reduce nitrate loss. Controlled drainage keeps the water table high during the off-season when crops are not growing. The high water table increases the rate of denitrification (a process that converts nitrate to harmless nitrogen gas (N_2) as soon as the saturated soil warms up in the spring) and reduces nitrate loss to the environment.

Controlled drainage can be combined with subirrigation to improve yields while protecting water quality. Subirrigation is irrigation back through the subsurface drain tiles. Subirrigation may be economical when fields are relatively level and need to be drained anyway, since additional infrastructure consists mainly of increased numbers of tiles the pumping system. One system being developed in Ohio combines a wetland for water treatment and a pond serving as a reservoir for subirrigation with a drainage system. This system has been shown to increase yields and reduce water quality impacts of drainage, although it is costly.

• Rehabilitation of abandoned jhum lands and measures to prevent soil erosion

Shifting cultivation, also known as 'jhum'cultivation in Northeast India is an ancient method of agriculture that is still practiced by tribal communities in many parts of the world, particularly in the wet tropics. In 1984, the Central Forestry Commission estimated that 6.7 million ha of cultivable area was affected by jhum in India. The continuance of jhum in the state is closely linked to ecological, socio-economic, cultural and land tenure systems of tribal communities. Since the community owns the lands the village council or elders divide the jhum land among families for their subsistence on a rotational basis. In this chapter, we take a close look at jhum cultivation from the point of view of ecological sustainability and tribal livelihoods, examine the role of agro-forestry, sericulture and horticulture as alternatives/supplementary activities and review the current thinking on methods to upgrade and develop jhum lands.

Land use and cropping pattern

The dry broadcast or 'punghul' method involves sowing in the month of March/April and harvesting in August/September. Wet sowing or 'pamphel' is done in the month of May/June and harvested during October/November. Transplanted paddy or 'aringba' is also sown in the month of May/June and harvested in the month of October/November. In the hilly areas, shifting cultivation is widely practiced, with settled terrace farming in foothill or low slope areas, above the adjacent rivers and streams. Depending on the slope, wet broadcast on bunded fields or dry broadcast on unbunded fields is practised. In the plains wet paddy rice cultivation is prevalent. Traditional methods of production are still widely used by the farmers, especially on the hill slopes. The land use/land cover patterns of the state have been classified into five categories, namely settlement, agricultural land, forest, water bodies and others. The last category of others includes rivers/streams, roads, water logged areas converted to new agricultural land, etc.

Characteristics of jhum cultivation

The characteristics of jhum cultivation are as follows: (i) Cutting and clearing of forest areas and burning of the dried biomass by fire, (ii) rotation of jhum land every four to seven years, (iii) use of human labour as the chief input, (iv) non-employment of animal implements or machinery, (v) collective ownership of land, (vi) reciprocal labour sharing and (vii) mixed cropping system. Women predominate in seed selection and planting, weeding, and other operations, while operations such as cutting of the jungle, clearing, burning of the cut under growth, etc., are done by men. Both men and women participate in harvesting. The produce is transported from the jhum land to the village by head-loading.

Based on a long-term study of jhum, that the version found in this region is that the cultivation is carried out on slopes of 30-40° angles; the climate is monsoonic with a high rainfall of over 2,200 mm followed by a dry winter and a brief warm summer, supporting a mixed sub-tropical humid forest. The normal jhum cycle is of three-four years, but rarely longer; and the forest is clear felled before planting'. Jhum enables multiple cropping of several crops which provide a balanced diet and also offers some form of crop insurance to the Jhumias in the event of failure of some crops. This system of food production might have worked well in the past when a balance was maintained between population and soil fertility as a result of a longer fallow cycle of 20 to 30 years. The shortening of the jhum cycle to an average of 45 years due to increase in population has resulted in a number of distortions appearing in the system including decreasing soil fertility and crop yields and inadequate management of fallows.

Impact of shifting cultivation

So long as the jhum cycle has duration of 10 years or more this type of cultivation did not pose any threat to the ecological stability and soils of the largely forested hill area. Underlying the view that shifting cultivation has alone been responsible for deforestation and environmental degradation is a deep-seated resentment felt by some sections of Manipuri society and the state administration at the lack of control over land and forests in the hills, the acknowledged power of traditional village councils and headmen, and the assertion by the hill tribes of their rights over local natural resources. The solutions offered in terms of regulating jhum cultivation include: a) resettlement of jhum farmers by relocation of their villages and provision of alternate means of livelihood; b) introducing terrace cultivation or forestry cooperatives on jhum lands; c) diversification into horticulture, floriculture and plantation crops and d) a complete change in the land tenure system in the hills whereby community owned and operated holdings are replaced by individual holdings. If these programmes are implemented without massive public investment and instituting a more participatory model of development planning in consultation with the communities for whom they are designed, they are, likely to cause major disruptions in the agricultural systems, food security, and way of life of the hill people and possibly endanger the very identity of tribal societies and their traditional institutions.

Recommendations for jhum control

The planning department has laid considerable emphasis on the control of shifting cultivation. The government has introduced certain measures aimed at i) restricting juming like allowing natural forest to grow in jhum lands, ii) initiating resource surveys, iii) increasing the area under terrace cultivation, iv) promoting programmes for intensive valley development and development of horticulture, v) plantation farming in jhum land and vi) the development of sericulture and a few forest-based industries. However, the performance of these programmes and schemes was very poor.

The primary cause of the failure to attract people to settled farming was that the new settlements provided were not readily accepted by the people due to their close attachment to their traditional villages and way of life. Moreover, the switch to terrace cultivation or the use of bullocks for ploughing causes great technical difficulties for jhum farmers. Limited availability of land for terrace farming at higher altitudes is also a major problem. Some terraces, built with retaining walls at a height of three-seven feet, are very difficult to maintain during the rains, resulting in the collapse of whole terraces with standing crops. A pre requisite for the success of terrace farming is the development of a proper technology for water management and water conservation. Since terraces are above perennial streams and rivers, it is essential that adequate power is available for lift irrigation schemes. Thus, while promoting terrace cultivation on the model recommended by the Indian Council of Agriculture Research (ICAR), due consideration and attention needs to be given to all the above-mentioned requirements.

Essentially, jhum control will only be possible if there are alternative land use systems. Alternative land use systems like horticulture require marketing and other infrastructural facilities to the farming communities. Until such systems are evolved any attempt to stop jhum cultivation will amount to depriving the people of their basic means of survival.

Along with development of land up to 50 per cent slope by constructing bench terrace, contour bund in the low hill slopes for permanent cultivation and creation of irrigation facilities by constructing dams, water harvesting structures, channels, etc, horticulture development along with soil and conservation plantation crop is vital. Afforestation too must be encouraged in the hills and non-agricultural activities to be given utmost priority, with technical, financial and marketing supports from the government and NGOs.

238   Rainfed Agriculture

Emerging strategies for jhum cultivation

a) Transfer of indigenous technology from one tribe to another or one area to another.

b) Upgrading of jhum by introducing variations in the species composition in the crop mix in order to increase economic returns and improve ecological efficiency.

c) Use of bamboo varieties and other fast-growing native trees as windbreaks to check wind and minimize the loss or water borne soil, ash and nutrients.

d) Introduction of appropriate rural technology such as rainwater harvesting, tanks, mini-hydels, biogas, to strengthen village ecosystems.

e) Participatory development based on jhum fallow management including community involvement in the selection of preferred species for fallow cycles.

f) Creation of the right kind of institutions for natural resource management atthe local level that are built on the pattern of the traditional institutions that already exist.

g) Redevelopment of valley wet rice cultivation and improvement of other landuse systems such as homestead gardens to increase productivity in the plains for state level food security.

Land rehabilitation

Land rehabilitation is an intervention designed to make a geo-ecological improvement. In most contexts, this involves the mitigation or reversal of land degradation caused by poor land husbandry practices, especially agricultural practices. The key issues in land rehabilitation concern to what degree the land should be rehabilitated to self-sustaining natural control and to what degree to a sustainable economic after-use, where future land quality is sustained by careful management and repair.

In many contexts, land rehabilitation works involve countering the physical symptoms of land degradation, which include losses of soil quality due to soil or subsoil compaction, and also accelerated runoff and erosion on hill slopes and in watercourses. In other contexts, the interventions involve the mitigation of wind erosion and pollution– especially by salts or industrial wastes, or military ordinance. The success of a land rehabilitation strategy, if it is not expressed in economic terms, is evaluated in the same terms as progress in ecological succession, commonly the integration, efficiency andresiliency of the geo-ecological system.

The driving force for land rehabilitation is land degradation. Land degradation is a composite term indicating the aggregate diminution of the productive potential of the land, including its major uses (rainfed arable, irrigated, rangeland, forestry), its farming systems (*e.g.* small holder subsistence) and its value as an economic resource. This term, which includes the subset of "desertification" refers to the decline of the biological productive potential of land, namely the entire geo-ecological system that includes soils, waters, climate, vegetation, topography and land use.

Land degradation may be an inherent property of the natural system (e.g. some of the eroded ravine lands of South Asia may be due to tectonic uplift generated channel incision). In other cases, climatic change may be implicated. More usually, land degradation is caused by a mismatch between the land's self-sustainable biological potential, its quality in human terms, and the way the land is used. Simply, the way the land is used causes more damage to the land than its restorative systems can compensate. This chapter deals with the physical and technical processes of land rehabilitation but both land degradation and rehabilitation are driven by social and economic causes. Ultimately, land degradation and land rehabilitation are human processes that reflect the ways in which human societies use and value the land that feeds them.

Afforestation: In a state like Manipur, where about 90% of the total area is in the hills and where forest is the biggest land use pattern, taking up large scale afforestation needs no emphasis. The scheme of afforestation has been taken up with the objectives of protecting land against erosion, restoration of degraded land to productive areas, better moisture conservation for improving productivity; reduce siltation in reservoirs and finally generation of employment opportunities.

- To protect the land against soil erosion.

- Restoration of degraded/abandoned jhum land and underutilized land to productive management.

- To construct soil conservation measures to check erosion and improve soil fertility and moisture content.

- Terracing, leveling, contour bunding and construction of small scale engineering structures like vegetative check dams, boulder sausage, bamboo spur, etc.

- To check siltation of rivers and increase the life spans of important dams, reservoirs etc.

- To improve the socio-economic condition of people.

- To generate employment opportunities.

Rehabilitation of Jhumias: To wean away the jhumias from the practice of shifting cultivation through an integrated approach involving agriculture, forestry, horticulture, animal husbandry *etc*.

- To divert the villagers from jhuming to settled farming.

- To provide jhumias alternative means of sustenance.

- To conserve the fertility and moisture content of soil.

- To generate employment.

- To create basic infrastructure facilities and develop villages.

- To create awareness among villagers.

Forest protection and fire control

- Patrolling of forest areas.

- Cutting of forest fire lines and digging of cattle-proof trenches.

- Opening of check posts at vulnerable points.

- Development of information network and social fencing.

- Construction of watch towers at strategic points.

- Purchase of fire extinguishers, fire alarms and fire suppressing instruments.

- Observation of "Forest Protection Day" every year in each district.

- Developing an immediate rewarding system for public informers who help in prevention of forest offence leading to confiscation of articles and conviction of the culprits.

Social forestry

The social forestry has been drawn up with the objective of meeting the fuelwood, fodder and small timber requirements of the local people from the nearby areas and thus leave the forests in the interior hill areas for maintaining the ecological balance. For implementation of the scheme, suitable barren and degraded forest area and wastelands near district headquarters and major villages in the hill areas are given due consideration. The scheme was taken up for the fuelwood and fodder plantation and execution of small scale engineering structures in selected areas.

Drainage Considerations and Agronomic Management; Rehabilitation 241

- To ensure adequate supply of fuel wood for cooking and heating purposes.
- To facilitate the supply of bamboo and small timber for agricultural and commercial purposes.
- To supply fodder for cattle.
- To generate employment of villagers.
- To reclaim degraded forests and wastelands.
- To conserve soil and water particularly in hill areas.
- To reduce the damage caused by shifting cultivation.
- Free distribution of seedlings to public and institutions.
- To generate a massive people's movement for environmental conservation.

Improving shifting cultivation practices on steep slopes: options and associated research priorities

1. Nutrient supplementation: Experimental increases in soil nutrient availability to plants in this region have enhanced crop growth and increased economic yields although often to a lesser extent. Enhanced yields per unit area could encourage farmers to burn and cultivate smaller plots, thereby extending the mean fallow period across the region.

2. Water supplementation: Soil water availability in this region may restrict crop production either directly as a result of plant water stress or indirectly as a constraint on soil nutrient supply to plants. Temporary small mobile or even permanent large rainwater harvesting tanks could significantly enhance yields.

3. Optimising crop choice: A farmer's choice of crop depends on environmental, farm management, economic and market access factors. Farmers are well aware that multicropping agroforestry practices promote stable productivity, as well as resilience against disease, drought years, and variability in crop market values. In essence, this cultivation approach creates a multi-layered canopy and a diverse rooting system that optimize overall light, nutrient and water use efficiency. Furthermore, farmers adapt their mix, tending to grow proportionally more tubers and rhizomatous crops and less cereal when using shorter fallow periods, or in relatively infertile soils. Finally, incorporation of wild or relatively unknown plant species into agricultural production may be beneficial.

4. Extending the site use period: Multiple cropping over successive years using appropriate choices of perennials along side annuals may be a very promising option for extending the duration of cultivation on a burned site. For

242   Rainfed Agriculture

example, the total regional area burned each year could be reduced to 1/3 of current rates if crop yields on a burned site were maintained at similar levels in the second and third years after burning. Starchy tubers such as cassava (*Manihot esculenta*), taro (*Colocasia esculenta*), and sweet potato (*Ipomoea batatas*), fruits such as banana and papaya (*Carica papaya*) and vegetables such as chillies, beans and squashes (*Cucurbita* spp.) could be grown during the mid-late growth phase of the first year's cereals. These crops will provide a protective soil cover when and after the cereals are harvested and will be ready for harvesting in the subsequent years. Further more, these crops restrict declines in soil fertility by promoting a protective vegetation and surface litter layer that restricts nutrient and particulate losses and enhances soil organic matter. Finally, crop yields have been successfully maintained on annually burned slopes in Nagaland for 6 years by using simple fences to restrict soil erosion losses.

5. Management to enhance the fallow recovery rate: The minimum fallow period required for sufficient natural regeneration of the vegetation and restoration of soil fertility to permit sustainable shifting cultivation is at least 10 years in North East India. Techniques such as plantation withnitrogen-fixing species to enhance fertility and protect against erosion may lower the minimum fallow period and therefore improve long term overall crop productivity per unit area of land.

6. Improved management of fire and its environmental impacts: Since substantial land areas appear to be burned but then not cultivated, there may be considerable potential for improving burning efficiency and there by reducing the total area burned each year and the associated detrimental environmental impacts. In other words, land burning should be restricted to the actual area required for cultivation in the following spring. Furthermore, since 30–50% of the nutrients in ash (mainly phosphorus and base cations) may be blown off burned sites by the strong winds that typically occur in North-East India prior to the monsoons, any measures to restrict ash loss are also likely to be beneficial.

Replacing shifting cultivation practices on steep slopes: options and associated research priorities

1. Nitrogen-fixing shrub hedge rows: Sloping agricultural land technology (SALT) is a form of continuous cultivation that was pioneered on gentle slopes in the Phillipines, and has been successfully adopted in many hilly tropical regions. Nitrogen-fixing shrubs are planted as dense hedge rows along slope contours and a diverserange of crops is cultivated in the inter-row areas.The hedgerows prolong fertility by restricting soil erosion and by replenishing soil nitrogen. Farm trials strongly suggest that this option has considerable potential in Mizoram

2. Slope terracing: Hand labourers or machinery canbe used to dig terraces that reduce downflow soil erosional losses and help to maintain soil moisture. Terracing initiatives in North-East India have apparently been unsuccessful in the past because their construction, maintenance and heavy fertilizer requirements meant that once government subsidies were removed, farmers reverted back to shifting cultivation. However, improved construction and management techniques suggest this option has substantial promise. For example, contour fencing can be used to promote initial terrace build-up, and to prolong terrace structure by resisting erosion.

3. Agro-forestry with anti-erosion plants: Bamboo, banana and many trees grow very well on steep slopes in this region and could substantially restrict soil erosion losses as part of an agro forestry multi cropping system.

4. Bamboo forest harvesting: Bamboo is particularly abundant across Mizoram, and could perhaps be harvested in a sustainable and ecological sensible way as a valuable crop. Although it has a low fuel energy per unit mass, itsabundance means that it has potential as a bio fuelin a region where energy resources are particularly scarce. Bamboo can also be processed into paper, as is done in the neighbouring state of Assam and has a wide range of other commercial values. Finally, bamboo may have potential as a substrate for biochar (pyrolysed biomass) that seems to enhance soil fertility and crop production as wellas soil carbon sequestration. Teak (*Tectona grandis*) also grows well in this region, implying that even scattered plantations may have considerable potential because of the wood's high market value.

12

Stress Physiology: Strategies for Mitigating Stress in Dryland Areas

Both dryland crops and irrigated crops experience an assortment of ecological or environmental stresses which include abiotic viz., drought, water logging, salinity, extremes of temperature, changes in atmospheric gases and biotic viz., insects, birds, other pests, weeds, pathogens. Dryland crops are more vulnerable to stress in the present context of changing climatic scenarios. The ability to tolerate or adapt effectively by challenging these stresses is a very complicated phenomenon would be due to various plant interactions occurred in the specific environment. Both biotic and abiotic stresses occur at various stages of plant growth and development and frequently more than one stress concomitantly affects the crop. It is difficult to distinguish effects of these stress factors on the performance of crop plants with respect to yield and quality of harvested products. This is of special significance of maximizing productivity of dryland crops in changing climate scenarios, with complex consequences for ecologically and environmentally sound Indian agriculture. In order to successfully meet this challenge, one should understand the various aspects of stresses in view of the current development and to promote a competitive dryland production system.

Abiotic stress

Abiotic stress is defined as the adverse effect of naturally occurring or non-living factors on the living organisms in a specific environment. The non-living variable must influence the environment to adversely affect the population performance or individual physiology of the plant in a significant way. This type of stress is primarily unavoidable and is the most harmful factor concerning the growth and productivity of crops in dryland areas.

Ex. Drought stress, cold stress, salinity, water logging, wind, extremes of temperature, high variability in radiation, changes in atmospheric gases

Biotic stress

Biotic stress is the stress that occurs as a result of damage done to plants by other living organisms, such as bacteria, viruses, fungi, parasites, beneficial and harmful insects, weeds, and cultivated or native plants. The types of biotic stresses forced on a plant for their growth and development depend on its ability to resist particular stresses under particular environment. Among the biotic stress, the majority of plant diseases are caused by fungi and pests.

A. Effects of abiotic stress on dryland crops

As said above abiotic stress occurs naturally and agronomists can think of mitigation strategies for this stress. While biotic stress are generally caused by pest and diseases, understanding the physiology is of less importance unless the better productivity is obtained in dryland areas. Hence, more discussion is required on abiotic stress which influencing growth and development of dryland crops in this chapter.

Soil moisture is the most limiting factor for getting profitable yields in dry farming and dryland farming situations. Understanding of the physiological processes that happen during moisture stress is needed to ameliorate the stress effects either by management practices agronomically or by altering the cropping pattern of the regions. Stress is the consequence of action of external factors on an organism/plants. Moisture stress specifies the action of shortage of water or excess of water on plants.

Causes

- Whenever transpiration surpasses absorption of water by the plants leads to stress

- After irrigation or rainfall, water deficit develops due to higher transpiration than absorption especially on hot middays. In such situations, stomata are open for a short period in the morning and evening and partially closed during the rest of the day.

- Whenever soil moisture reaches about -15 bars, plants show wilting symptoms (permanent wilting point) and at -60 bars, plants die permanently (ultimate wilting point).

1. Effect of moisture/drought stress on physiology of plants

a. Water relations: When the moisture is limiting in dryland areas, had negative impacts on water relations in plants such as reduction in leaf water potential and relative water content mainly due to transpirational demands. In addition to

Stress Physiology: Strategies for Mitigating Stress in Dryland Areas 247

this, continuous decline in turgor of the plants likely anticipated. In such plants, leaf wilting is a common symptom could be due to turgor loss. Increase in leaf and canopy temperature might have hindered the plant growth. Turgidity of guard cells decrease with increasing moisture stress in plants.

B. Photosynthesis and Respiration

Drought stress induces several changes in various physiological, biochemical, and molecular components of photosynthesis. Photosynthesis is reduced due to reduction in photosynthetic rate, chlorophyll content, leaf area and increase in assimilate saturation in leaves. Biochemical reduction of CO_2 by altering several biochemical process will effect under moisture stress. Reduction in leaf expansion, tillering or branching and increase in leaf senescence in moisture stressed plants is a collective phenomenon. Translocation of assimilates is affected. The photosynthesis apparatus, Photosystem II is comparatively more tolerant to drought stress than heat stress. Respiration increases with mild stress and if the stress continues, water content and respiration will be affected considerably. Moisture stress coupled with high temperature can result in decreases in leaf and root respiration in the short term.

Drought can influence photosynthesis either by stomatal closure and decreasing flow of CO_2 into mesophyll tissue or by directly impairing metabolic activities. The main metabolic changes are declines in regeneration of ribulose bisphosphate (RuBP) carboxylase/oxygenase enzyme. In general, during the initial onset of drought stress, decreased conductance through stomata is the primary cause of decline in photosynthesis. At later stages with increasing severity, drought stress causes tissue dehydration, leading to metabolic impairment.

C_3 plants

C_3 plants (dryland pulses, rice, wheat etc) are closely associated with lower productivity at given water-use than C_4 plants mainly because of RuBP carboxylase which is insensitive to atmospheric CO_2 concentrations. Atmospheric CO_2 concentrations are strongly inhibitive to CO_2 uptake in C_3 plants and fixed directly by RuBP carboxylase. Photosynthesis is smaller in C_3 plants leads to lower crop water use efficiency (WUE) for a given rate of transpiration. Photosynthesis decrease in drought conditions because of stomatal limitations. The drought-sensitive metabolism of the C_3 plants have greater recovery of photosynthesis upon irrigation, through increased stomatal conductance.

C_4 plants

The C_4 plants having "Kranz leaf anatomy" are intimately higher productivity at given water-use. The C_4 crops are maize, sorghum, pearl millet, and various forage grasses is essentially a pumping mechanism that moves CO_2 from the mesophyll cells and causes high CO_2 concentrations in the vascular-bundle sheath cells by using CO_2 from decarboxylated C_4 acids in the mesophyll cells. This results in a high utilization efficiency of low intercellular CO_2 concentrations due to the PEP carboxylase enzyme in the C_4 plant. Photosynthesis is greater in C_4 than in C_3 plants has advantage translated into a greater WUE. Hence, C_4 plants has more drought resistance capacity and resulted in relatively better yield under stress. However, under well-watered conditions C_4 plant had greater WUE leads to better economic returns.

c. Anatomical changes

Drought stress from early growth stages leads to decrease in size of the cells, intercellular spaces, thicker cell wall and greater development of mechanical tissues. Hardening of the cell wall and its reduced extensibility will diminish and even it stops cell growth. Size of epidermal cells are reduced without reduction in the number of stomata per unit number of cells. Stomata per unit leaf area tend to increase under drought stress.

d. Metabolic reactions

Severe water deficit in dryland areas causes decrease in enzymatic activity in the plants. Amount of enzymes involved in hydrolysis remain or increase while peroxidase activity decrease. Accumulation of sugars and amino acids takes place under moisture stress conditions. Proline, an amino acid accumulates under moisture stress.

e. Hormonal relationships

Cytokine, gibberlic acid and IAA, growth promoting hormones produced in the plants were tend to decrease under moisture stressed environments. Contrastingly, growth regulating hormones (ABA, ethylene, betain) will produce and tendency to increase by the moisture stressed plants. ABA known as "stress hormone", acts as water deficit sensor to minimize the loss of tissue water potential. ABA is endogenous hormone also caused dormancy and found in wilting leaves and subsequently induced stomatal closure.

ABA effects and consequences in the plant

- General growth decreases

- Cell division and cell expansion decrease

- Germination and tillering decrease

- Root growth and root hydraulic conductance increase

- Flower abscission increase

- Pollen viability decrease

- Grain and fruit growth decrease

- Starch synthesis in cereal grains increase

- Plant reserve mobilization to the grain increases

- Leaf senescence and dormancy increase

f. Protoplasmic dehydration

Dehydration is severe, protoplasm become rigid and brittle. Tissues become desiccated, protoplasm becomes increasingly dense and its viscosity gradually increases.

g. Growth processes and nutrition in plants

The expansion of cells and cell division are reduced under stress resulted in decreased growth of leaves, stem and fruits. Moisture stress during sowing will severely affect germination and hence leaf area, and leaf expansion and root development. The primary processes involved in the plant growth and development are cell division and cell elongation. Generally, cell division is appraised to be less sensitive to moisture stress than cell enlargement. However, comparatively mild water stress influenced considerably on both cell expansion and cell division, even before photosynthesis or respiration is affected. Under moisture stress conditions, maintenance of cell turgor plays an important role in cell growth. Comparatively, water stress during early stage of the cell division and cell expansion were able to recover fully than at the final phase of cell growth and division. Decreased leaf senescence and loss of leaf area are often termed as tolerance and avoidance mechanism during stress.

- Leaf area expansion is most sensitive growth process and decreased drastically under limited drought stress, mainly because of smaller cells and reduction in the number of cells produced by leaf meristems.

- Leaf area is reduced under stress to moisture. Leaf area is reduced in old leaves by reducing mature cell size, whereas, in younger leaves, inhibition of cell division resulted in fewer cells per leaf.

The general effects

1. Reduction in leaf numbers, rate of expansion, and final leaf size during mild stress.

2. Under severe stress, the rate of leaf elongation decreases and leaf growth can cease.

3. Total leaf area can be decreased under drought stress.

4. Continued drought stress can accelerate leaf senescence and lead to death of leaf tissue, resulting in leaf drop, particularly old and mature leaves.

5. Rewatering plants after a relatively short period of stress (3–5 days) does not completely eliminate the effects of drought on the senescence process. In contrast to drought, heat stress can stimulate cell division and cell elongation rates.

6. Drought stress often decreases stem growth in response to changes in internal water status.

Moisture stress affects fixation, uptake and assimilation of nitrogen in dryland pulses. Nitrogen fixation by leguminous plants reduced due to reduction in leghaemoglobin in nodules and number of nodules. NPK uptake is reduced by moisture stress. Nitrogen assimilation is affected due to reduction in nitrate reductase activity. Nitrogen deficiency under moisture stressed conditions will increase stomatal resistance. Phosphorus deficiency also increases in stressed plants.

h. Developmental and reproductive processes

In general, if stress occurs before flowering, the duration of the crop increases. If the stress occur after flowering, the duration of the crop decreases. Panicle initiation is most sensitive stage to moisture stress due to reduction in cell expansion. Anthesis is another important moisture sensitive stage as moisture stress causes drying of pollen and loss of viability of pollen. Grain development decrease due to reduction in leaf area and photosynthesis.

By the intensity and duration of stress, duration of the developmental stage as well as transition of one developmental stage to other is hampered. Development is mostly influenced by photoperiod (daylength) or vernalization is mainly a function of temperature depenedent. Both developmental rate of leaves (individual organs) and the advancement of the entire plant are quantitatively reliant on temperature. For example, high temperature encourages more rapid development of leaf canopy and also causes the overall crop development rate to increase so that the crop growing season is shortened.

Some of the developmental process as influenced by high temperature and subsequently moisture stress are as follows;

1. The length of time from floral initiation (panicle initiation) to anthesis (panicle exsertion) is decreased by moderate drought and/or temperature stress but is increased by severe stress.

2. Drought stress during panicle development inhibits the conversion of vegetative to reproductive phase and plants remain vegetative until the stress is relieved.

3. Panicle initiation in sorghum was delayed by as many as 2 to 25 d and flowering by 1 to 59 d under drought stress, with more severe effects when drought was imposed both at early and late stage of panicle development.

4. Drought and heat stress can delay the panicle initiation but also can cause the cessation of panicle development at any stages between panicle initiation and flowering.

5. Severe drought or heat stress inhibits panicle exsertion and also delays flowering.

6. Drought stress or heat stress during flowering and anthesis inhibits pollen development and causes sterility because of decreasing pollen function.

7. Drought and/or heat stress also shortens the spike development duration (period during which potential kernel or seed numbers are determined) and the grain-filling duration (during which the grain or seed weight are determined).

8. Drought stress during later stages of panicle or flower development decreases seed numbers and can also increase the duration from seed-set to full seed growth.

9. For cereal crops, longer periods of vegetative and reproductive development are often necessary to improve reproductive potential (number of productive tillers and kernels) and also leaves and tillers to provide assimilate supply during the grain filling.

10. Both drought and heat stress decreases the seed-filling duration, leading to smaller seed size in rice and legumes such as groundnut, soybean.

2. Effect of heat stress on physiology of plants

At insignificant persistent temperatures, have direct positive influence on rates of development (e.g., seed germination rate and flowering) and occurs most

252   Rainfed Agriculture

rapidly. At constant soil moisture conditions, percentage seed germination increases with increasing temperature above base temperature and later on decreases at higher temperature. Soil moisture stress coupled with high temperature may greatly affect the growth and development of crop plants in dryland areas. High temperatures generally increase leaf appearance rates and leaf-elongation rates, while decreasing leaf-elongation duration. Heat stress resulted in significant increases in leaf numbers, particularly when reproductive development was arrested without any decrease in leaf photosynthetic rates.

a. Growth and reproductive development

High temperatures at early stages resulted in reduced germination percentage, plant emergence, abnormal seedlings, poor seedling vigor, reduced radicle and plumule growth of geminated seedlings. High temperature causes loss of cell water content for which the cell size and ultimately the growth is reduced. Reduction in net assimilation rate (NAR) is also another reason for reduced relative growth rate (RGR) under high temperature in C_4 crops. The reproductive tissues are the most sensitive to high temperature during flowering time leads to loss of entire grain crop cycles.

- High temperature stress also leads to changes in morphological indications like scorching and sunburns of leaves and twigs, branches and stems, leaf senescence and abscission, shoot and root growth inhibition, fruit discoloration and damage.

- Reduced number of tillers with promoted shoot elongation, leaf area and productive tillers/plant were drastically reduced under HT (30/25 °C, day/night). High temperatures may alter the total phenological duration by reducing the life period.

- Severe heat stress decreases stem growth, decreased plant height, decreased root growth by reduced root number as well as root length and root diameter.

- Significant decrease in floral buds and flowers abortion, in some cases may not produces flowers, fruit or seed.

- High night temperatures (32 °C) increase spikelet sterility resulting from excessive ethylene production was also noticed.

b. Photosynthesis and respiration

Photosynthesis is one of the most heat sensitive physiological processes in plants. The processes of photosynthesis are more tolerant to heat stress and are mostly stable in the temperature range of up to 30 to 35°C, depending on crop species.

However, very high temperatures (>40°C) can negatively affect photosynthesis. The thermal effects of photosynthesis and respiration are related to membrane function and membrane integrity. In general, heat stress influences membrane fluidity, induces membrane leakiness, and influences the integrity of protein and membranes.

At high temperature, photosynthesis decreases due to decreasing the solubility of oxygen resulting in increased photorespiration. The activity of Rubisco are also decreased at high temperatures by inhibiting the enzyme Rubisco activase. Photosystem II is most sensitive to heat stress and greatly reduced. In general, photosynthesis is temporally (only during daytime) and spatially (only in green tissues) restricted, while respiration occurs continuously and in all organs. Respiration exponentially increases with increasing temperatures from 0 to 35 or 40°C, reaching plateau at 40 to 50°C. At temperature above 50°C, respiration decreases because of damage to respiratory mechanism.

- Major changes occur in chloroplasts by causing injury of thylakoids, loss of grana stacking and swelling of grana under heat stress.

- Closure of stomata under heat stress is another reason for decreased photosynthesis that affects the intercellular CO_2.

- Soluble proteins and Rubisco binding proteins (RBP) are reduced under heat stress which resulted in hampered photosynthesis.

- Starch and sucrose synthesis is greatly affected by reduced activity of sucrose phosphate synthase, ADP-glucose pyrophosphorylase, and invertase at high temperature.

- Reduced leaf water potential, reduced leaf area and pre-mature leaf senescence which have negative impacts on total photosynthesis performance of plant under heat stress.

- Accumulation of heat shock proteins (HSPs) is a mechanism of stress resistance to heat which protect the plant tissues.

3. Effects of salinity on plants

Soil salinity is a major factor that limits the yield of agricultural crops, threatening the capability of crop production system. At low salt concentrations, yields are mildly affected or not affected in all the crops. While at high concentrations most of the crop plants, will not grow in high concentrations of sodium salts (100-200 mM NaCl). While the plants which tolerates to high salt concentration (halophytes) mostly grown in coastal and arid regions can survive in excess of 300-400 mM due to specific mechanisms of phylogenetic adaptation.

High salinity affects plants in the following ways:

- High concentrations of salts in the soil disturb the capacity of roots to extract water (osmotic effect), and

- High concentrations of salts within the plant itself can be toxic due to inhibition of many physiological and biochemical processes such as nutrient uptake and assimilation and hence growth of plants reduced.

- Old leaves in the sensitive plant die and reduce the photosynthetic capacity of the plant. This exerts an additional effect on growth.

- Water use by plants is decreased due to reduced shoot growth by reduction in the leaf area development relative to root growth.

Prominent changes occur in plants under salinity/sodicity stress are mentioned below:

1. High Na^+ transport to shoot

2. Preferential accumulation of Na in older leaves

3. High Cl^- uptake by the plants

4. Lower K^+ uptake by the plants

5. Lower fresh and dry weight of shoots and roots

6. Higher partitioning of dry matter to sensing leaves

7. Low P and Zn uptake

8. Poor root growth.

Synthesis of Heat Shock Proteins (HSPs) under high temperature stress

Synthesis of HSPs or stress proteins, is a tolerance mechanism occuring in plants. Synthesis of HSPs is extensive response to diversity of environmental stress such as drought, salinity, heavy metal stress and oxidative stress. These are highly conserved and produced in all plants to protect from the damage caused in plants from high temperature/heat stress as well as other stresses. Heat shock proteins acts as a defence mechanism against high temperature that leads to changes in the cellular membrane structure, protein metabolism, level of enzymes and photosynthesis activity. Hence these are called "Heat-shock proteins or Stress-induced proteins or Stress proteins". Heat stress/high temperature, affects the metabolism and structure of plants, especially cell membranes and many basic physiological processes such as photosynthesis, respiration, and water relations.

B. Effects of biotic stress on dryland crops

Biotic stress is stress that occurs as a result of damage through to plants by other living organisms viz., bacteria, viruses, fungi, parasites, beneficial and harmful insects and weeds. The difference between biotic and abiotic stresses are discussed earlier. Among the biotic stress, the major stress which occurred economic loss of plants under dryland regions are pest and diseases and the intensity and occurrence depends on climate and topography. Losses caused by biotic stress remains high and farmers may face greater difficulty in controlling biotic stresses compared to abiotic stress. The biotic stress injury mainly influence on crop yields that influences population dynamics, plant development and ecosystem nutrient cycling.

Losses due to pests and disease in crop plants remain to pose a significant threat to crop production and ultimately food security. Now a days, intensive farming systems became increasingly dependent on synthetic chemical pesticides for controlling pests and diseases and became unsustainable. Application of more and more pesticides leads to development of resistance in the target pests, and has negative impacts on biodiversity. More importantly, plants will have increased susceptibility to pest due to the implications of climate change. Under such situations, gaining better knowledge on physiology of plants could leads to sustainable control of biotic stress. For plant biotic stress resistance and process to manipulate, we require a detailed knowledge of these interactions at a wide range of scales.

Effect of biotic stress on plant growth

Both insect and diseases will affect photosynthesis, as chewing insects reduce leaf area and virus infections reduce the rate of photosynthesis per leaf area. Vascular- wilt fungi greatly affects the water transport and photosynthesis by inducing stomata closure in the plants.

Response of plants to biotic stress

Selection of wide range of plants and suitable crop varieties defences against pathogens and pests which act to minimize frequency and impact of attack. It include both physical and chemical adaptations. Utilization of high metal ion concentrations derived from the soil permit plants to reduce the harmful effects of biotic stressors (pathogens, herbivores etc.) called induced resistance. Meanwhile preventing the severe metal toxicity by way of safeguarding metal ion distribution throughout the plant with protective physiological pathways. At the same time, successful pests and pathogens have evolved mechanisms to overcome both constitutive and induced resistance in their particular host species.

256   Rainfed Agriculture

Mitigation of abiotic stress by crop management

Moisture or drought stress management is already discussed in the Chapter 7. Apart from adoption and resistance mechanisms, we need to know other mitigation strategies for maximizing productivity and profitability of the dryland farmers sustainably by preserving resource base of the dryland soils.

1. Conservation tillage

The conservation tillage system is designed to conserve soil moisture from one season to another or from one year to the next. Conserving soil moisture is virtuously, depending on climate and crops of the regions. Increasing storage of soil moisture by with or without conservation tillage is typical moisture conservation practice in dryland farming. The benefit of conservation tillage in terms of increasing available soil moisture to the crop depends on soil water-holding capacity, climate, topography and management practices. Percent increase in soil moisture availability to the crop but the amount of soil moisture is not remarkable under probability of crop failure and success. The conservation tillage carries additional benefits such as improved soil nutrients availability and the eradication of certain soil-born pests, such as nematodes.

Conservation tillage is not a novel concept or practice, which has recently gained wider and sometimes enthusiastic acceptance. The principles involve, minimized tillage operations to conserve soil structure and to maintain ground cover by mulch, such as stubbles which reduce water runoff and increase soil infiltration. Rotation of crops under dryland areas is not suitable because of soil moisture constraints. Intercropping with cover crops/ pulses could increase the soil moisture availability by harvesting runoff and utilized for next season crop. .

Deep tillage is a system to overcome hardpan, very high bulk density and compacted soils can be performed by deep plowing or deep ripping. Deep plowing involves actual plowing to depth which is an expensive operation and is uncommon in dryland farming. Deep ripping is somewhat less expensive and more often used in crop production. The important consideration in deep ripping is to operate at the correct depth in order to break the hardpan, no less and no more.

2. Furrow ditches/dikes

These techniques constitute a field surface tillage manipulation to minimize runoff away from the field and provides more time to infiltration. Short basins were made as furrows by small dikes and is very amenable to grow row crops such as cotton, corn and sorghum. It is generally considered effective for increasing rainfall capture and raising dryland crop yields where annual rainfall ranges between 500 and 800 mm.

3. Water harvesting and spreading

As discussed in the Chapter 4, water harvesting is broad technique to conserve moisture in dry land areas in minimizing abiotic stress. For more detailed information in this important practice see Water harvesting and moisture conservation Chapter.

4. Diversification of farming

Diversification of farming is an ancient but an effective approach to reduce the risk associated with farming in unpredictable environments. Reduced diversification to the extent of mono-cropping is possible only with a high level of control like irrigation over the crop environmental conditions. Diversification of cropping to reduce risk is especially important under dryland conditions. Crop diversification takes an advantage where crops grown in a single stress environment. Crops differ in their response to a given environment and this difference is used to reduce the risk associated with growing one crop. "Mixed cropping" or "intercropping" is an example of a traditional and a successful approach to crop diversification on a single piece of land, where two or more crops are grown together in various possible patterns. Planting of various crop varieties offer a better chance of reducing loss due to environmental stress, as compared with growing one variety only. Varietal diversification is constructed mostly on differential phenology, primarily flowering dates when two or more environmental stress likely occurs.

13

Tools and Implements Used in Dryland Agriculture

In dryland areas, moisture conservation accounts prime most important concept for maximizing productivity of dryland crops. Varied topographic and agro-climatic conditions ranging from arid to semi-arid climates with high evaporation because of varied temperature regimes permit the cultivation of drought resistance crops. Reducing the cost of cultivation without compromising the socio-economic status of farmers is prime importance. Conservation of moisture by reducing the weed menace is great task performed by the dryland farmers. Under such situations, effective utilization of dryland tools and implements used are of a primitive nature throughout the dryland farmers for higher factor productivity. Traditional farm tools and implements for self-sustenance have been adapted through experience over generations to meet emerging socio-economic and farming challenges.

Relatively small yields consequently larger areas must be farmed for a given return and the successful exploration of dryfarming requires the adoption of methods. The methods which enable farmers to do the maximum effective work with the smallest expenditure of energy. The type of soils and topographic conditions largely influence the type, size and shape of particular tillage tools/implements among the dryland farmers. The following is a list of local tools/implements found in the various dryland regions.

Dryland implements

A. Primary and secondary tillage implements

1. Plough: Tillage is the basic operation in farming done to create favourable

260   Rainfed Agriculture

conditions for proper seed placement, germination and plant growth. This is done mainly with a plough by turning the soil to a depth of seven to ten inches is a fundamental operation of dry-farming. Farmers have been using plough since old ages. The basic components of the plough are a shoe, a share, a body, a handle and a beam. Ploughs used in dry lands have shoes which are generally of a triangular section. The primary purpose of ploughing is to turn over the upper layer of the soil, bringing fresh nutrients to the surface, while burying weeds and the remains of previous crops and allowing them to break down. Plowing and cultivating a soil homogenizes and modifies the upper 12 to 25 cm of the soil to form a plow layer.

a. Traditional wooden plough or country plough

Country plough is being used and chosen in most part of the dryland regions where small farmers and farm owners of scattered lands are unable to use tractors. It can work on gravely soils with stones and other obstacles or soils with a greater proportion of sand particles which come across during the sequence of ploughing. Indigenous plough is an implement which made of wood with an iron share point. It is drawn with bullock and drudgery will be a problem on the animals. It cuts a 'V' shaped furrow and opens the soil but there is no inversion. Ploughing operation is also not perfect because some unploughed strip is always left between furrows. This is reduced by cross ploughing, but even then small squares remain unploughed. It covers an area of 0.15 to 0.25 ha in eight hours depending on the type of soil and moisture content. Present day, these wooden ploughs were replaced by modern plough for more suitable for the proper turning and stirring of the soil.

b. Mouldboard (MB) plough

The mouldboard plough is considered as, the most acceptable plough in dryland regions because it possess short rapid curving and it pulverizes the soil most thoroughly. It is highly preferred plough since saves cost on production from the resulting saving of energy. The parts of MB plough are body, mouldboard or wing, share, landslide, connecting rod, bracket and handle. This type of plough leaves no unploughed land as the furrow slices are cut clean and inverted to one side resulting in better pulverization. Both animal drawn and tractor drawn MB ploughs are available and these ploughed to a depth of 25 to 30 cm. These are used where soil inversion is necessary. To grow crops regularly in less fertile areas, the soil must be turned to bring nutrients to the surface. The mouldboard plough greatly reduced the amount of time needed to prepare a field, and as a consequence, it improves the efficiency of farmers by allowing to work on larger area of land. The ploughing pattern of low and high ridges in the soil forms water channels, allowing the water to drain. The efficiency of

MB plough is more and can be ploughed about 4.0 ha per day in light soils and about 1 ha in heavy soils.

Advantages of mouldboard ploughing

1. It aerates the soil by loosening it.

2. It incorporates crop residues, solid manures and commercial fertilizers.

3. It reduces nitrogen losses by denitrification, accelerates mineralization and increases short-term nitrogen availability for transformation of organic matter into humus.

4. It controls many perennial weeds and pushes back the growth of other weeds.

Disadvantages

1. Ploughing leaves very little crop residue on the surface, which otherwise could reduce both wind and water erosion.

2. Overploughing can lead to the formation of hardpan.

3. Soil erosion due to improper land and plough utilization is possible.

Table 1: List of tools/implements in dryland regions

Operation	Tools/implements used
Tillage and bed/land preparation tools Leveller	Wooden Plough, Country plough, Cultivator,
Interculture operation tools	Spade, Harrow, Hand Hoe
Harvesting	Sickle, Plough
Post-harvest	Winnower, Basket/Gunny bags, Large sieve
Miscellaneous	Chisel, Saw, Iron Hammer

2. Ridge ploughs or bund former : These are the simple implements for making bunds or ridges to conserve soil moisture. The plough has two mould boards, one for turning the soil to the right and another to the left. These are mounted on a common body. The ridge plough is also used for earthing up of crops. These are used to make broadbed and furrows by attaching two ridge ploughs on a frame at 150 cm spacing between them. Bunds are made along the contour to prevent soil erosion during heavy rains on steep slopes.

3. Cultivator: Tractor drawn cultivator is an implement used by medium and large farmers in dryland regions. It is used for finer operations like breaking clods and to bring soil to a fine tilth in the preparation of seedbed. It is used to further loosen the previously ploughed land before sowing to destroy weeds

262   Rainfed Agriculture

that germinate after ploughing. Cultivator has tynes with staggering rows to provide clods and plant residues can freely pass through the tynes without blocking. The number of tynes ranges from 7 to 13. Cultivators are often similar in form to chisel ploughs, but their goals are different. Cultivator teeth work near the surface, usually for weed control, whereas chisel plough shanks work deep beneath the surface. Consequently, cultivating also takes much less power per shank than does chisel ploughing.

Advantages of cultivators include,

1. Destroy the weeds in the field.

2. Aerate the soil for proper growth of crops.

3. Conserve moisture by preparing mulch on the surface.

4. To sow seeds when it is provided with sowing attachments.

5. To prevent surface evaporation and encourage rapid infiltration of rain water into the soil.

4. Harrows: These are the secondary tillage implements used for shallow cultivation in operation of seedbed, crushing of clods, covering seeds, destroying weed, intercultivation and harvesting of some crops. Disc, tyne and blade harrows are commonly used in dryland regions. Blade harrows used for breaking soil crust after rains, for summer deep harrowing to control weeds and smoothing out the surface of the soil. Indigenous blade harrows are bullock drawn implement have the disadvantage of clogging and lack of penetration on hard soil. The soil and clods do not pass through the blade harrow where weeds are more. Under such circumstances several types of iron blade harrows have been developed. Harrows differ from cultivators in the way that they disturb the whole surface of the soil, such as to prepare a seedbed, instead of disturbing only narrow trails that skirt crop rows (to kill weeds).

B. Leveller and roller

The leveller is made of any locally available wood having beam of 2 m length and shafts are generally made of bamboo sticks are used for levelling land. Wooden made levellers are commonly used in dryland areas from which most of the clods are crushed due to its weight. Rollers are used mainly to crush the hard clods and to compact the soil in seed rows.

The most feasible idea of levelling is to level the field to its best condition with minimal earth movement and then differ the water supply for the field condition. This viewpoint is most economic because land levelling is expensive and large earth movements may leave significant areas of the field without fertile topsoil. Uniquely, farmers attach leveller to cultivator and do ploughing along with levelling after primary tillage. This is cost saving and more economical to farmers.

Benefits of land levelling

1. Better crop establishment

2. Optimization of water use efficiency

3. Improves the efficiency of water, labour and energy resources utilization.

C. Sowing implements

It has already proven that proper sowing is one of the most significant operations of the dry-farming. The unfashionable method of broadcasting is prevailed in some parts of dryland regions. The drawback of broadcasting method is that, neither the quantity of seed used nor the manner of placing the seed in the ground can be synchronized. Seed dibbling in the furrows behind the plough is another method of sowing in dryland areas. This is more laborious and time consuming method. Economically, seed drill are developed for dryland areas where labours are scarce.

1. Seed drill

Seed drill consists of a wooden beam to which 3 to 6 tynes are fixed and these tynes open the furrows into which the seeds are dropped. Holes are made into these tynes and into these holes, metal seed tubes or bamboo are fitted at the bottom ends. The seed is forced to fed into tubes at a uniform rates by skilled labours so placed as to enable the seed to fall into the furrows in the ground. Leveller or chain dragging behind the drill and covering seed furrow quite completely. In this method of sowing, it is very desirable that the soil should be pressed carefully around the seed to ensure better germination. The seed drill is already a very useful implement and is rapidly being made and used to meet the special requirements of the dryland farmers. Crops based seed drill are developed and being used and are very assisted to the dry land farms.

2. Fertilizer cum seed drill

With the knowledge of seed drill, fertilizer cum seed drills are developed where fertilizers are placed at a depth of 5 cm and 5 cm away from seed rows for effective utilization of fertilizers. Both operations like dibbling seeds and fertilizers are done simultaneously. It is similar to seed drill, but extra tynes and hopper for drilling fertilizers are fitted. Mechanical seed drill also developed for enhancing

264   Rainfed Agriculture

efficiency with automated drilling of seeds and fertilizers having rotating disc which has holes and bottom plates coincides, seed and fertilizers falls into the tube and then to soil. The distance between two holes in rotating disc is proportional to the inter-row spacing of crop.

D. Interculture operation tools

The operations performed in the field after sowing but before harvesting the crop are called as intercultural operations. It is done in between the crop rows by breaking the upper surface of soil. Advantage of intercultivation operation is that, uprooting the weeds (unwanted plants), aerating the soil, thereby promoting the activities of microorganism and making good mulch, so that moisture is conserved inside the field by reducing the evaporation.

In general, primary and secondary tillage equipment viz., country plough, blade harrows, ridge ploughs and weeders are used for earthing up and to control weeds in widely spaced crops. Functions of these equipments are discussed above. Apart from these tools, hand operated intercultural tools are also widely used in dryland farmers.

1. Spade or Hand hoe: The most popular manually operated weeding tool used in the farm is hand hoe or spade. It is utilized for doing operations such as formation of bunds, ridges and furrows, and irrigation channels. It consists of an iron blade and a wooden handle. The operator holds the handle and cuts the soil with the blade to a shallow depth of 2-3 cm thereby weeds are cut and soil is stirred.

2. Wheel hoe or Junior hoe: The wheel hoe is a widely accepted weeding tool for weeding and intercultural opperations in rows of standing crops. It comprises of a wheel, tool frame, a set of replaceable tools and a handle. It is a long handled tool operated by pushes and pull action. Different types of hoes such as straight blade, V -blade, sweep, shovel, etc. can be used for weeding, soil mulching, stirring etc. Long handle reduces drudgery to operator and wheel reduces energy requirement for pushing. Efficiency of junior hoe is more compared to wheel hoe.

E. Harvesting tools

The most common type of harvesting implements used in dryland areas are sickles. A sickle is a hand-held agricultural tool with a variously curved blade typically used for harvesting grain crops (rice, wheat, maize, sorghum, barley, pulses and millets) or cutting succulent forage or grasses chiefly for feeding livestock. Serrated sickle was designed as 'C' shaped/curved and it is preferred more than other tools and implements in view to ease the harvesting operation. With the help of sickle the ear heads, branches or even whole plant could be harvested.

F. Postharvest tools and implements

1. Winnower: Wind winnowing is a primeval cultures for separating grains from chaff. In threshing, loosening of grain or seeds from the husks and straw, is the step in the chaff-removal process that comes before winnowing. Techniques included using a winnowing fan (a shaped basket shaken to raise the chaff) or using a tool (a winnowing fork or shovel) on a pile of harvested grain. In its simplest form it involves throwing the mixture into the air so that the wind blows away the lighter chaff, while the heavier grains fall back down for recovery.

2. Sieve: This is used for the separation of different types of grains for elimination of foreign materials.

G. Miscellaneous tools

Hammer, jumper, wedge and shovel and hand saw are also used from time to time in the various farm operations.

Despite of their widespread use, even today, these indigenous implement/tools in general are not agronomically sound and as a result lower the efficiency and increase tiredness of the operator. Though modern tools and implements have been developed for various activities, the usage by the dryland farmers is negligible. The socio-economic status of the farmers plays a key role in utilizing the modern tools and implements. Custom hiring serves are available at all the village level agricultural departments for carrying all the field operations at less cost.

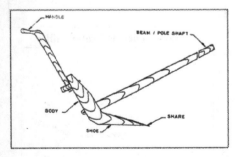
Country plough

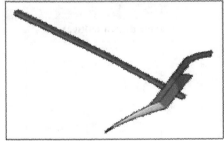
Wooden plough

266   Rainfed Agriculture

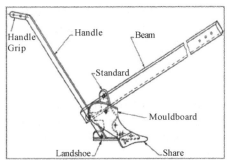

Iron plough parts

Animal drawn iron plough

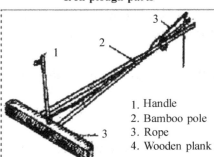

1. Handle
2. Bamboo pole
3. Rope
4. Wooden plank

Animal drawn leveller

Tractor drawn leveller

Tractor drawn cultivator

Tractor operated rotary tiller

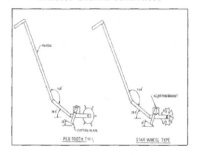

Improved dryland weeders like

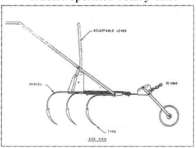

Wheel hoe and Junior hoe

Tools and Implements Used in Dryland Agriculture 267

Tractor drawn disc harrow

Hand held weeder- Junior hoe

Animal drawn seed drill

Tractor drawn seed cum ferti drill

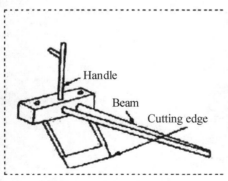
Animal drawn blade harrow

Tractor drawn blade harrow

Spade

Hand hoe

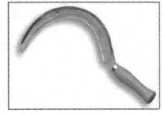
Sickle

14

Collection of Biometric Data on Dryland Crops and Its Interpretation

---

Understanding the principles involved in plant growth and development is prime important for managing agronomic practices for higher productivity of dryland crops sustainably. Systematic study of plant growth based on the variability and interpreting plant development based on biometric methods is called as growth analysis. Plant data in turn biometric data is collected from the field and analysed over the period to know the growth condition of the plants. Based on biometric data, agronomic practices like application of fertilizers, irrigation and weed management practices were employed. Several growth indices are devised for identifying environmental or genetic factors as sources of variation in plants. Both whole plant and their component organs (leaf) are basis for calculating growth indices.

The most common parameters used in growth analysis are leaf area index, crop growth rate, relative growth rate, net assimilation rate and leaf area duration.

1. Leaf area: Leaf area is significant for photosynthesis. In plants, as growth proceeds the leaf area will increase by increasing the rate of production of assimilating material. Thus, its estimation indicates both assimilating area and crop growth. It is measured by destructive sampling where leaves are separated from plant for dry matter and used to measure leaf area by manual methods. Leaf area is calculated from the green leaves collected from the field by taking length× breadth× number of leaves then multiplied by k factor.

- Horizontal leaves (sun-flower, cotton, pulses), have large k values, habitually 0.7–1.0,

- Erect leaves (cereals, grasses) will have small k values, habitually 0.3–0.6.

2. Leaf area index (LAI)

Leaf area index is the ratio between leaf area to the ground area. For crop production leaf area per unit land area is more important than leaf area of individual plants. Maximum LAI is attained just prior to flowering in cereal crops. As the senescence proceeds crops lose leaves and hence LAI declines during grain filling or maturity. It denotes the canopy photosynthetic active radiation (PAR) interception and interception increased rapidly during early stages of growth. Excessive leaf area development can be counter-productive and economic yield may be reduced due to self-shading and resource allocation to leaf production. The leaf area index is calculated at different crop stages as per the formula given by Watson (1952).

$$\text{Leaf area index} = \frac{\text{Total leaf area, cm}^2}{\text{Total land area, cm}^2}$$

Example 1. Leaf area of groundnut at 30, 60 and 90 days are given in the table. Calculate the LAI of the spacing is 30x15 cm

DAS	Leaf area (cm^2/plant)
30	352.8
60	1821.0
90	1949.9

Solution

Area occupied by groundnut plant = 30x15 cm^2
= 450 cm^2

Leaf area of one plant at 30 DAS = 380 cm^2

LAI at 30 DAS = $\dfrac{352.8}{450}$

= 0.784

LAI at 60 DAS = $\dfrac{1821.0}{450}$

= 4.05

LAI at 90 DAS = $\dfrac{1949.9}{450}$

= 4.33

Note

- LAI<1, means reduced PAR interception due to insufficient leaf area
- LAI=1, leaf area at plant density is sufûcient for radiation interception and plant dry matter production
- LAI=4, means 4 layers of leaf area is covered same land area in which, incident PAR was intercepted completely (about 90%), may cause self-shading.

3. Leaf area duration (LAD)

Leaf area duration measures the ability to produce leaf area on unit area of land throughout its life. Yield of dry matter is a function of leaf area, net assimilation rate and duration of leaf area. It was calculated at various days period as per the formula given by Powar *et al.* (1967).

$$LAD = \frac{LAI_1 + LAI_2}{2} \times (t_2 - t_1)$$

Where, LAI = Leaf area index; t_2-t_1 = time periods

Example 2. Calculate the LAD when LAI of clusterbean at 30, 60 and 90 days are 0.784, 4.051 and 4.332 respectively.

Solution

LAI of clusterbean plant at 30 DAS = 0.784

LAI of clusterbean plant at 60 DAS = 4.051

LAI of clusterbean plant at 90 DAS = 4.332

$$LAD = \frac{LAI_1 + LAI_2}{2} \times (t_2\text{-}t_1)$$

$$LAD \text{ for 30 to 60 DAS} = \frac{LAI_1 + LAI_2}{2} \times (t_2\text{-}t_1)$$

$$= \frac{0.784 + 4.051}{2} \times (60\text{-}30)$$

$$= \frac{4.835}{2} \times 30$$

$$= \quad 2.4175 \times 30$$

$$= \quad 72.5 \text{ days}$$

$$\text{LAD for 60 to 90 DAS} \quad = \quad \frac{LAI_1 + LAI_2}{2} \times (t_2\text{-}t_1)$$

$$= \quad \frac{4.051 + 4.332}{2} \times (90\text{-}60)$$

$$= \quad \frac{8.383}{2} \times 30$$

$$= 4.1915 \times 30$$

$$= 25.7 \text{ days}$$

Note

- Low LAD at early growth stages could lead to incomplete development of sinks and thus lower yield.

- 125.7 days means, the leaf area developed between 60 to 90 days period at 30 days duration under best management conditions, could produce same leaf area in 127.7 days under normal conditions.

4. Absolute growth rate (AGR)

AGR is defined as growth rate of plant at any given time. It indicates at what rate the crop is growing i.e. whether the crop is growing at a faster rate or slower rate than normal. It is expressed as gram of dry matter produced per day calculated as per the formula given by Radford (1967).

$$AGR = \frac{W_2 - W_1}{t_2 - t_1} \text{ g plant}^{-1} \text{ day}^{-1}$$

Where, W_1 and W_2 are dry weights of plant at time t_1 and t_2, respectively

Example 3 : Total plant dry weight of groundnut at 30, 60 and 90 days are given in the table. Calculate the AGR.

DAS	Total dry weight (g/plant)
30	2.70
60	19.22
90	27.58

Solution

Dry weight of groundnut plant at 30 DAS $=$ 2.70 g/plant

Dry weight of groundnut plant at 60 DAS $=$ 19.22 g/plant

$$\text{AGR} = \frac{W_2 - W_1}{t_2 - t_1} \text{ g plant}^{-1}\text{day}^{-1}$$

$$\text{AGR for 30 to 60 DAS} = \frac{19.22 - 2.70}{60 - 30}$$

$$= \frac{16.52}{30}$$

$$= 0.55 \text{ g plant}^{-1}\text{day}^{-1}$$

$$\text{AGR for 60 to 90 DAS} = \frac{27.58 - 19.22}{60 - 30}$$

$$= \frac{8.36}{30}$$

$$= 0.28 \text{ g plant}^{-1}\text{day}^{-1}$$

5. Relative growth rate (RGR)

RGR is defined as the rate of increment in dry weight per unit plant weight per unit of time. This parameter indicates rate of growth per unit dry matter produced. The yield of a field crops is the weight per unit area of harvested produce. In RGR growth could be considered as a process of continuous compound interest, where in interest (efficiency production) is also added to the principle (dry weight gain) to calculate interest. The dry weight of a plant is not all productive capital since a considerable part is leaf-supporting material and is not active in photosynthesis. Any dry matter increase is due largely to photosynthesis of light intercepting tissue so a good measure of the productive capital of the plant

274 Rainfed Agriculture

is leaf area. Hence, RGR is an index of plant efficiency. It is expressed as gram of dry matter produced by a gram of dry matter produced in a day and it is calculated as per the formula given by Blackman (1919).

$$RGR = \frac{\log_e W_2 - \log_e W_1}{t_2 - t_1} \text{ g g}^{-1}\text{day}^{-1}$$

Where, w_1 and w_2 are dry weights of plant at time t_1 and t_2, respectively. $\log_e$, natural logarithm.

Example 4. Total plant dry weight of groundnut at 30, 60 and 90 days are given in the table. Calculate the RGR.

DAS	Total dry weight (g/plant)
30	2.70
60	19.22
90	27.58

Solution:

Dry weight of groundnut plant at 30 DAS $= 2.70$ g/plant

Dry weight of groundnut plant at 60 DAS $= 19.22$ g/plant

$$RGR = \frac{\log_e W_2 - \log_e W_1}{t_2 - t_1} \text{ g g}^{-1}\text{day}^{-1}$$

Natural log value for 2.70 is 0.99 and for 19.22 is 2.96

$$\text{RGR for 30 to 60 DAS} = \frac{2.96 - 0.99}{60 - 30}$$

$$= \frac{1.97}{30}$$

$$= 0.0657 \text{ g g}^{-1}\text{day}^{-1}$$

Natural log value for 19.22 is 2.96 and for 27.58 is 3.32

$$\text{RGR for 60 to 90 DAS} = \frac{3.32 - 2.96}{60 - 30}$$

Collection of Biometric Data on Dryland Crops and Its Interpretation   275

$$= \frac{0.36}{30}$$

$$= 0.012 \text{ g plant}^{-1}\text{day}^{-1}$$

Note

1. RGR is a measure used to quantify the speed of plant growth.

2. RGR nearly always decreases over time as the biomass of a plant increases. Could be due to

   - Non-photosynthetic biomass (roots and stems) increases,

   - The top leaves of a plant begin to shade lower leaves and

   - Soil nutrients can become limiting.

6. Crop growth rate (CGR)

CGR is the gain in dry matter production on a unit land in a unit of time. It simply indicates the change in dry weight over a period of time. It is expressed as gram of dry matter produced per m^2 per day.

$$\text{CGR} = \frac{1}{A} \times \frac{W_2 - W_1}{t_2 - t_1} \text{ g m}^{-2}\text{day}^{-1}$$

Where, W_1 and W_2 are dry weights of plant at time t_1 and t_2, respectively; A is land area

Example 5. Total plant dry weight of groundnut at 30, 60 and 90 days are given in the table. Calculate the CGR where the spacing of groundnut is 30x 15 cm

DAS	Total dry weight (g/plant)
30	2.70
60	19.22
90	27.58

Solution

Dry weight of groundnut plant at 30 DAS $= 2.70 \text{ g/plant}$

Dry weight of groundnut plant at 60 DAS $= 19.22 \text{ g/plant}$

$$\text{CGR} \qquad = \frac{W_2 - W_1}{t_2 - t_1 \times A} \text{ g m}^{-2} \text{ day}^{-1}$$

$$\text{CGR for 30 to 60 DAS} = \frac{19.22\text{-}2.70}{60\text{-}30 \times (0.30 \times 0.15)}$$

$$= \frac{16.52}{30 \times 0.045}$$

$$= \frac{16.52}{1.35}$$

$$= 12.23 \text{ g m}^{-2} \text{ day}^{-1}$$

$$\text{CGR for 60 to 90 DAS} = \frac{27.58\text{-}19.22}{60\text{-}30 \times (0.30 \times 0.15)}$$

$$= \frac{8.36}{30 \times 0.045}$$

$$= \frac{8.36}{1.35}$$

$$= 6.19 \text{ g m}^{-2} \text{ day}^{-1}$$

7. Net assimilation rate (NAR)

It indirectly indicates the rate of net photosynthesis. NAR can be expressed on basis of leaf weight (including respiration of the total plant) and chlorophyll content, as well as leaf area. The progress of dry matter accumulation and mean net assimilation rate can be designated by yield. It is expressed as g of dry matter produced per m^2 of leaf area in a day. For calculating NAR, leaf area of individual plants has to be used but not leaf area index. It was calculated at 0-30, 31-60, 61-90 and 91-120 days period as per the formula given by Gregory (1917).

$$\text{NAR} = \frac{(W_2\text{-}W_1)(\text{Log}_e L_2 - \text{Log}_e L_1)}{(t_2\text{-}t_1) \ (L_2\text{-}L_1)} \text{ g m}^{-2} \text{ day}^{-1}$$

Where L_1 and W_1 are leaf area and dry weight of plants at time t_1, and L_2 and W_2 are leaf area and dry weight of plants at time t_2.

Collection of Biometric Data on Dryland Crops and Its Interpretation 277

Example 6. Leaf area and total plant dry weight of groundnut at 30, 60 and 90 days are given in the table. Calculate the NAR

DAS	Leaf area (cm²/plant)	Total dry weight (g/plant)
30	45.37	2.70
60	361.84	19.22
90	475.82	27.58

Solution

Dry weight groundnut plant at 30 DAS = 2.70 g/plant

Dry weight groundnut plant at 60 DAS = 19.22 g/plant

Leaf area of groundnut plant at 30 DAS = 45.37 cm²/plant

Leaf area of groundnut plant at 60 DAS = 361.84 cm²/plant

$$\text{NAR} = \frac{(W_2\text{-}W_1)(\text{Log}_e L_2 - \text{Log}_e L_1)}{(t_2\text{-}t_1)\,(L_2\text{-}L_1)} \ \text{g m}^{-2} \ \text{day}^{-1}$$

Natural log value for 45.37 is 3.81 and for 361.84 is 5.89

$$\text{NAR for 30 to 60 DAS} = \frac{(19.22\text{-}2.70)\,(5.89\text{-}3.81)}{60\text{-}30 \times (361.84\text{-}45.37)}$$

$$= \frac{16.52 \times 2.08}{30 \times 316.47}$$

$$= \frac{34.36}{9494.1}$$

$$= 0.0036 \ \text{g m}^{-2} \ \text{day}^{-1}$$

Natural log value for 361.84 is 5.89 and for 475.82 is 6.17

$$\text{NAR for 60 to 90 DAS} = \frac{(27.58\text{-}19.22)\,(6.17\text{-}5.89)}{60\text{-}30 \times (475.82\text{-}361.84)}$$

$$= \frac{8.36 \times 0.28}{30 \times 113.98}$$

$$= \frac{29.86}{3419.4}$$

$$= 0.0087 \text{ g m}^{-2} \text{ day}^{-1}$$

Limitation

- When using only dry weights, NAI may not give true treatment effects on root-shoot ratio.

- Photosynthesis may take place in organs other than leaf lamina, namely stems and inflorescence, and using leaf area alone may then lead to underestimates of the photosynthetic area.

8. Relative water content

The water content of tissues was expressed relative to their fully turgid water content when floated on water as per the method of Barrs and Weatherley (1962) at any growth stages. Relative water content (RWC) also termed as 'relative turgidity' because when the water content was measured by floating excised tissue to attain the fully hydrated condition indicates that the turgor of osmotically adjusted cells. Likewise, water contents would become higher than normal. RWC has two advantages.

- RWC does not vary along a leaf, because uptake of water related to the dry weight of leaf.

- It is an indirect measure of the change in turgor of leaf because any changes in RWC are proportional to changes in turgor of a leaf.

1.0 g leaflets selected randomly from each treatment and fresh weight was measured accurately on electronic balance. The weighed leaflets were floated in petriplate containing distilled water and allowed to imbibe water for four hours. After four hours, leaflets were blotted gently and weighed for recording turgid weight. After taking turgid weight, the leaflets were dried in an oven at 65°C for 48 hours and the dry weight was recorded. The RWC was calculated by using the following formula.

$$\text{RWC (\%)} = \frac{\text{Fresh weight (g) - Dry weight (g)}}{\text{Turgid weight (g) - Dry weight (g)}} \; 100$$

Collection of Biometric Data on Dryland Crops and Its Interpretation 279

Example 7. Leaf dry weight, turgid weight and dry turgid weight of clusterbean at 30, 60 and 90 days are given in the table. Calculate the RWC when 1.0 g fresh weight of plant samples taken.

DAS	Leaf dry weight	Turgid weight	Dry turgid weight
30	0.23	1.08	0.22
60	0.32	1.26	0.30
90	0.40	1.19	0.50

Solution

$$RWC\ (\%) = \frac{\text{Fresh weight (g) - Dry weight (g)}}{\text{Turgid weight (g) - Turgid dry weight (g)}} \times 100$$

$$RWC\ (\%)\ \text{for 30 DAS} = \frac{1.0 - 0.23}{1.08 - 0.22} \times 100$$

$$= \frac{0.77}{0.86} \times 100$$

$$= 89.5\ \%$$

$$RWC\ (\%)\ \text{for 60 DAS} = \frac{1.0 - 0.32}{1.26 - 0.30} \times 100$$

$$= \frac{0.68}{0.96} \times 100$$

$$= 70.8\ \%$$

$$RWC\ (\%)\ \text{for 90 DAS} = \frac{1.0 - 0.40}{1.08 - 0.22} \times 100$$

$$= \frac{0.60}{0.86} \times 100$$

$$= 69.8\ \%$$

Note

- Floating of leaf discs more than 4 h for rehydration, as respiration and carbohydrate metabolism can reduce the concentration of soluble sugars and therefore of osmotic pressure.

- If measuring the RWC of growing tissues, carry out the rehydration step at 4°C, or an appropriate temperature to inhibit growth.

- The WC varies along a leaf, being greater at the base and lower at the tip. This is not just due to effects of transpiration, it reflects the different anatomy of cells and tissues along the leaf. Either a whole leaf should be sampled, of a known age or stage of development, or a defined portion of the leaf should be taken, *e.g.* a segment in the middle.

- Unstressed leaves will have a RWC of 90-95% depending on the humidity (VPD) and light. Stressed and wilted leaves may drop as low as 50%. Few leaves can recover from a RWC of 40%. In the dark, the RWC will be about 99%.

15

Suggested Readings and Acknowledgement

For the preparation of this book, the authors have drawn heavily from the following sources.

Aamodt,O.S and Johnston W.A. 1936. Can.J.Res.14: 122-152.

Adams, J. E. 1962. Agron. J. 54: 257-261.

Barrs, H.D. and Weatherley, P.E. 1962. *Australian Journal of Biological Sciences*, **15:**413-428.

Blackman, V.H. 1919. *Annals of Botany*, **33:**353-360.

Chen, D; Sarid, A. and Katchalski, E.1968.Proc. Nat.Acad.Sci;US. 61: 1378-83.

De, G.C. 1989. Fundamentals of Agronomy. Oxford and 1 BH pulishing Co. Pvt. Ltd. New Delhi.

El Sharkawy, M. Loomsi, R.S. and Williams, W.A. 1968. J. Applied Ecol.5:241-51.

El Sharkawy, M and Hesketh 1965. Crop Sci. 5: 517-21.

Eveneari, M. 1962. Arid Zone Research, 18:175-196.

Giri, G., Singh, R.R. and De, R. 1983. Indian J.agric Sci. 53:899-994.

Gregory, F.G. 1917. Third annual report, experimental and research station, Chesnut.

Grundbacher, F.J. 1963. Bot. Rev. 29:366-81.

Hatch, M.D. and Slack, C.R.1970. Ann. Rev. Pl. Physiol.21: 141-162.

Hatch, M.D., Slack, C.R. and Johsnon, H.S. 1967. Biochem.J. 102: 417-22.

Harrold et al. (1959)

Kassas, M.1966. In Arid lands: A Geographical Appraisal Hills, E. S. (ed), UNESCO, Paris pp. 145-80.

Killian, C. and Lemee, G. 1956. In Handbuch der Pfanzenphysiologie, Ruhland, W. (ed), Springer Verlag, Berlin, Band III. Pp. 607-726.

Kramar, P.J. 1959. In Plant physiology Vol.II. Steward, F.C. (ed), Academic press, New York. Pp. 607-726.

Levitt, J. 1958. Protoplasmatologia, 6:87.

Levitt, J.1972. Response of plants to environmental stress (ed) Academic press, New York.

Loomis, R.S., Williams, W.A. and Hall, A.E. 1971. Ann. Rev.Pl. Physiol. 22: 431-68.

282 Rainfed Agriculture

McDonough, W.T. and Gauch, H.G. 1959. Bull.Md. Agric. Exp. Sta. Pp.103.

Murata, Y. and Iyama, J. 1963. Proc. Crop Sci. Soc. Japan, 31: 315-22.

Oppenheimer, H.R. 1960. Arid zone res. (UNESCO, Paris) 15: 105-38.

Polunin, N. 1960. Introduction to plant Geography and some related sciences, Longmans Green, London and Barnes and Noble, New York.

Powar, J.P., Willis, W.O., Grunes, D.L. and Reichaman, G. 1967. *Agronomy Journal*, **59**: 231-234.

Radford, P.J. 1967. *Crop science*, **8**:71-175

Reddy and Redd:, 2002.

Prasad and Singh, 1994

Singh et al. 1990

Singh et. al. 1997

Salim, M.H., Todd, G.W. and Schlehuber, A.M. 1965. Ibid. 57:603-7.

Shields, L.M. 1958. In Morphology in relation to xerophytism (ed) New Mexico Highlands Univ. Bull. Pp. 15-22.

Shimshi, D. and Ephrat, Y. 1970. Report to Ford Foundation.

Srivastava, O.S. 1987. Policy issues of Agricultural Economics. Allied Publishers, New Delhi.

Stalefelt, M.G.1956. In Handbuch der Pfanzenphysiologie, Ruhland, W. (ed), Springer Verlag, Berlin, Band III. Pp. 351-426.

Steiner, J.L. Day, J.C.; Pupendick, R.I., Meyer, R.E. and Bertrand., A.R. 1988. Advances in Soil Science 8:79-122.

Stocker, O. 1960. Res.Rev. 15: 63-104.

Stutte, C.A. and Todd, G.W. 1967. Phyton 24: 67-75.

Subramanian, V.B. and Singh, R.P. 1983. Dryland Agriculture Research in India. Thrust in the Eightes All India Coordinated Research Project on Dry land Agriculture, Hydrabad.

Verma, B., Chinnamani, S., Bhola, S.N., Rao, D.H., Parsad, S.N. and Parkash, C. 1990. Twenty-five years Research on Soil Water and conservation research in ravines of Rajasthan CSWCRTI, Research Center, Kota, 1 215.

Watson, D.J. 1952. *Advances in Agronomy*, **4**: 101-145.

Went, F.W. 1965. In: Desert Research, Research Council of Isreal. Pp.232-7.

Wischmeier (1959)

16
Practical Excercises

Exercise No. 1. Rainfall analysis and interpretation

Procedure: Collect the rainfall data from the Department of Agrometeorology and interprete it zonewise and district wise for various districts of Haryana. Correlate it with last year's rainfall.

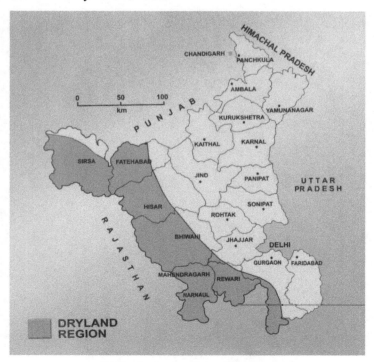

Fig.1 : Dryland regions of Haryana.

Interprete the data presented in the above figure both zonewise and districtwise.

284 Rainfed Agriculture

Table 1 : Distribution of annual rainfall according to seasons in India

Rainfall	Duration	% of annual rainfall
South west monsoon	June- September	74
Post monsoon	Oct- Dec	10
Winter (North-east Monsoon)	January- February	3
Pre- monsoon	March- May	13

Intreprete the distribution of annual rainfall according to seasons in India.

Exercise No. 2 : Implements used in dryland agriculture

The need to have improved implements for seeding, fertilization and interculture operations was long being felt in dryland areas.

Ridger seeder (Tractor drawn)

Bajra the principle *kharif* dryland crop suffers at germination stage due to problem of soil crusting/seed burying by rains after sowing. Ridge sowing of the crop could help to get rid of this problem. Also there was a definite problem to get proper germination of *rabi* oilseed crops raised on conserved moisture or as a second crop after short duration *kharif* pulse crop due to receding condition of soil moisture in the profile, ridger seeder is quite suitable implement to mitigate this problem. In *kharif* the ridger seeder place the seeds on top of the ridges in 30/60 cm paired row system and in rabi place the seeds deep in the moist zone and ensures proper coverage with moist soil.

Construction details

It consist of a tool bar holder for three point hitch mounting to hold a 2 inches square tool bar, 9 feet in length. Three ridgers provided on tool bar, form the ridging assembly. The wings of the ridgers are adjustable to form ridges and furrows of various dimensions. The 9 inch length of the tubing mounted to the ridger frame is shaped to direct the flow of fertilizer to the bottom of the furrow. Hoe type furrow openers for seed placement are mounted to the tool bar so that the openers are behind the fertilizer placing device and seed is placed above the fertilizer. The depth of seedling can be varied by adjusting the position of the hoe openers relative to the tool bar and ridger assembly. The furrow openers can also be adjusted to sow the seed in furrows or on the side of the ridges. A lugged gauge wheel is provided to regulate the depth of furrows and to drive the seed and fertilizer attachment.

A seed box fabricated from galvanized metal sheet is provided. Fluted roller type seed metering system to meter out the seeds ranging from chickpea to rapeseed is provided. A north western fertilizer attachment was mounted ahead of the seed box.

Multipurpose seed drill cum ridger seeder (Tractor drawn)

Farmers owing tractor in Haryana has drylands and irrigated lands sow wheat, cotton and other irrigated crops in his irrigated areas. The ridger seeder as such was not acceptable to the farmer due to its inability to sow wheat. To increase the versatility of ridger seeder as multipurpose seed drill cum ridger seeder was developed. Multipurpose seed drill cum ridger seeder has detachable arrangement which can be used to make the machinery as seed drill as well as ridger seeder. Recently, arrangement for sowing of intercropping system was also included in multipurpose seed drill cum ridger seeder.

Zero-till machine

It consists of a rectangular tool bar 200 X 60 cm. Seed and fertilizer boxes fabricated from galvanized metal sheet provided with fluted rollers each to meter out seed and fertilizer is provided. Nine combined furrow openers are provided to make it nine row seed drill. Two big clamps to fix two ridge bottoms on the main front tool bar are provided to make it ridger seeder. This implement was tested to sow bajra, cotton, moong, guar in *kharif* and mustard, wheat, gram in *rabi* and found quite promising. The wheat was sown at row spacing of 17 cm. The germination was found very well.

Ridger seeder (Bullock drawn)

Tractor drawn machinery helps in speedy work and timeliness of operation, but tractor power is not with in easy reach of small farmers. A simple attachment on country plough to enable it to sow the seeds on ridge as well as in furrow was designed. This implement was vigorously tested and some shortcomings were noted.

Interculture implements

Moisture is a limiting factor in raising a good crop. Weeds can not be allowed to sustain as these will compete with plants for nutrients as well as for moisture. The rapidity of interculture and weeding operation plays a vital role as timeliness of weeding is must. It was found that by breaking the soil crust and forming shallow soil mulch in field helps in maximum moisture conservation.

Bullock drawn blade hoe

It consists of a frame fabricated from mild steel flat size (40 X 10mm) in such a way that different size of blade can be fitted in the frame. Various blades of dimension 5 mm thick and 100 mm wide of 250 to 450 mm length fitted in the frame such that it makes a cutting angle of 15- 20 degree with the soil surface. The frame is provided with a 80 x 50 mm beam made of seasonal hard wood. The length of the beam is 2.5 metre.

Wheel hand hoe

It consists of a mild steel wheel having 400 m dia. The wheel is provided with eight spokes of 6 mm thick rod. Wheel hub with cast iron bush and mild steel axle is also provided A mild steel frame of variables sizes 150mm to 250 mm are mounted on the wheel. Two handless made ¾ inches dia 16 gauge steel pipe provided with 125 mm grips are attached on the frame. This implement tested and found labour saving having 1man/acre/day coverage.

Testing and evaluation of Modified Ridger Seeder for Pearlmillet and Raya sowing

Bed planter

Multi purpose seed cum fertilizer drill

Zero till drill

Laser land leveler

Practical Excercises 289

Exercise No. 3 : Agronomic measures of soil and moisture conservation.

Various method of moisture conservation in dryland areas are by storing in the root zone, by checking loss of water by evaporation and by water harvesting. Rainfall water can be conserved in following ways:

A) By storing more of rainfall in root zone

1. Tillage,
2. Contour farming of bunding,
3. Vertical mulching,
4. Subsurface barrier,
5. Addition of pond sediments of organic matter,
6. Addition of gypsum.

B) By checking loss of water through evaptripiration

1. Mulching,
2. Weed control,
3. Use of antitranspirants,
4. Crop thinning.

C) By water harvesting

1. Water harvesting *in situ*
   i) Inter row water harvesting
   ii) Inter plot water harvesting
2. Water harvesting for recycling

Elaborate above agronomic measures in your own words.

Exercise No. 4. Collection of biometric data on crops and its interpretation

The most common parameters used in growth analysis are leaf area index, crop growth rate, relative growth rate, net assimilation rate and leaf area duration.

Leaf area index (LAI)

Leaf area is important for photosynthesis. Its estimation indicates both assimilating area and growth. For crop production leaf area per unit land area is more important than leaf area of individual plants. Leaf area index is the ratio between leaf area to ground area.

LAI = Leaf area/ Ground area

Absolute growth rate (AGR)

It indicates at what rate the crop is growing i.e. whether the crop is growing at a faster rate or slower rate than normal. It is expressed as gram of dry matter produced per day.

$AGR = W_2 - W_1 / t_2 - t_1$

Where W_1 and W_2 are dry weights of plants at times t_1 and t_2 respectively.

Relative growth rate (RGR)

This parameter indicates rate of growth per unit dry matter. It is similar to compound interest, where in interest is also added to the principal to calculate interest. It is expressed as gram of dry matter produced by a gram of existing dry matter in a day.

$RGR = \log_e W_2 - \log_e W_1 / t_2 - t_1$

Where W_1 and W_2 are dry weights of plants at times t_1 and t_2 respectively.

Crop growth rate (CGR)

It is the rate of growth of crop per unit area and expressed as $g/m^2/day$.

$CGR = 1/p \times W_2 - W_1 / t_2 - t_1$, where p is land area.

Leaf area duration (LAD)

Yield of dry matter is a function of leaf area, net assimilation rate and duration of leaf area. Leaf area duration of a crop is a measure of its ability to produce leaf area on unit area of land throughout its life.

Measurement of plant water status

Plant water status can be measured by following methods

- Water content
- Relative water content
- Water saturation deficit.

Material required: plant leaves, water, petri dishes, forceps, oven, weighing balance

Procedure

- Select physiologically functional leaves
- Cut these leaves into different dishes of size one centimeter diameter.
- Take fresh weight of leaves as early as possible
- Put the leaves into petri-dishes for 4 to 5 hours
- Take turgid weight
- Put the leaves in to oven at 65^0 for 24 hours
- Note the dry weight.

Water content can be calculated as:

- Water content = Fresh weight – Dry weight / dry weight
- Relative water content = Fresh weight – Dry weight / Turgid weight – Dry weight
- Water saturation deficit = 100 – relative water content

Measurement of plant height

Material required: Scale

Procedure

- Select five representative plants from a plot in each environment
- Tag them
- Take the plant height measurement of plants weekly, 10 days, monthly interval with the help of meters rod from surface to growing tip of the plant
- Calculate the average plant height of plant by dividing the total value by five and express the result in cm per plant.

292   Rainfed Agriculture

Precaution: Never raise the leaf erect because it leads to error.

Measurement of dry weight of crop plant under different environment condition

Material required: Sickle or blade, paper or envelope, oven, electronic balance

Procedure

- Select five plants from each plot under particular environment.
- Cut the plant from the base of the stem.
- Air dry or sun dry the plant.
- Put the sample in the oven at temperature of 65°C.
- Take its weight with the help of electronic balance.
- Express the result in g/plant by averaging the total weight five plants.

Precaution

- Before putting the sample in oven it should be sun dried otherwise it would catch fire.
- While making the weight measurement, plants should be completely dried otherwise it may lead to error.
- Temperature of oven should be fixed at 65°C because if it will be more then plants will catch fire and if it is less then plant will not dry properly.

Measurement of plant growth in terms of number of tillers per branch or plant under different environments

Procedure

- Select the five representative plants of a crop in a particular environment
- Count the total tillers per branch
- Express the result in tillers /branch per plant by averaging the number of tillers or branches.

Precaution

Late mortality of some tillers gives error in results in term of growth.

Measurement of leaf area as growth in different environments

This is done by three methods

1. By multiplying length and width of leaf

2. Graph paper method

3. Leaf area meter.

Procedure

- Select the 5 plants from particular environment.
- Classify the leaves in to three classes i.e. large, medium and small.
- Count the number of leaves in each categories.
- Calculate the average length and breadth of each categories.
- Multiple the two dimensions and find out the area of each leaf.
- The leaf area of each category multiplied with its number and summation of leaf area of all the leaves will give total leaf area.
- Total area divided by five will give area per plant.

Graph paper methods

- Select the 5 plants from particular environment
- Classify the leaves in to three classes *i.e.* large, medium and small
- Put these leaves on the graph paper and take their impression on graph paper
- Count the number of leaves as
- ¾ square b) ½ squares c) ¼ squares d) Full squares
- Sum up the total area and then average the total area dividing by five gives area per plant.

Leaf area meter

Procedure

- Switch on the instrument
- First set the instrument on zero
- Put the samples on belt and it gives the area of leaf in the form of m^2 or cm^2

294   Rainfed Agriculture

Exercise No. 5 : Studies on mulches and anti-transpirants

Various types of mulches are as under

Mulching : With respect to dryland mulch is any material placed at soil surface with a view to conserve the soil water. These may be various types such as - Straw or residue mulch, soil mulch, plastic mulch, vertical mulch and chemical mulch.

a) Straw mulching can help in conserving soil moisture in following ways -

i) By checking loss of water from soil through process of evaporation because it reduces heating of soil from radiation and reducing wind speeds near soils surface.

ii) By reducing runoff and increasing infiltration of water.

iii) By checking weed growth.

b) Soil mulch: Surface mulch of dry soil 5-8 cm deep by obstructing the rise of water to the surface through capillary action effectively reduced loss of water as compared with effectively reduced loss of water as compared with a soil having an undisturbed surface.

c) Chemical mulch : It has been found that Hexadecanol, a long chain alcohol mixed with surface 1/4 inch of the soil reduced evaporation by 43%, this material which is resistant to microbial activity remained effective for more than a year. The surface layer of treated soil dried out more rapidly than that of untreated soil, creating a diffusional barrier to evaporation.

d) Vertical mulch : Vertical muching is a recent innovation which is found to be helpful in situ conservation of moisture for increased crop yields. This has been specially found to be suitable to black soils of Decan plateau whose intake rate are very low. It consist of jowar stubbles kept in trenches of 40 cm deep, 15 cm wide protruding 10 cm above ground level. Such trenches spaced at 4-5 meters increased crop yields by 400-500% in drought years and 40-50% in normal years ever control.

Plants transpire water vapours continuously from all above ground parts particularly through leaves. This process of evaporation of water from the aerial parts of plants is termed as transpiration. Approximately 99 percent of the water taken by the plants roots is transpired to the atmosphere. Transpiration occurs through different types of apertures such as cuticles, lenticels and stomata. Among these apertures stomata accounts for 90-97 per cent of transpiration. Transpiration is considered as a necessary evil. Reduction in transpiration may help in maintaining favorable water balance in dry farming. Any material that is applied to plant surfaces with the aim of reducing or inhibiting water loss from plant surface is called antitranspirants.

Scope of antitranspirants

- Under dryland area to reduce water losses through transpiration.
- In costly irrigation facilities for extending the irrigation intervals.
- For reducing transplanting shock of nursery plants.

Types of anti-transpirants: Based on mode of action, anti-transpirants are of four types

Stomata closing type: Most of the transpiration occurs through stomata on the leaf surface. Spray material used for various purposes such as certain fungicide (phenyl mercuric acetate), herbicide (atrazine) and metabolic inhibitors have been found to cause the closure of stomata and thereby reduction in transpiration. The effectiveness of an antitranspirants is also depends on the coverage of lower surface of leaves, interaction of the antitranspirants with external environments, surface anatomy of leaves and rate of formation and growth of new leaves.

Film forming type: Foliar spray of waxy or plastic emulsions such as mobileaf, hexadecanol and silicone produce an external physical barrier outside the stomatal opening to retard the escape of water vapour through stomatal opening. The film so formed should have more resistance to the passage of water than to that of carbon dioxide. Film type antitranspirants which provide selective type of permeability barriers to water vapours and carbon dioxide diffusion in the required directions have not yet been found so far.

Reflectance type: White reflecting materials such as whitewash or kaoline spray form a coating on the leaves and increase the leaf reflectance (albedo). By reflecting the large amount of radiation, they reduce leaf temperature and vapour pressure gradient from leaf to atmosphere and thus reduce transpiration. Application of 5% kaolin spray has been found to reduce transpiration losses markedly.

Growth retardants: Foliar application of chemicals such as cycocel reduces shoot growth, increases root growth and induces stomatal closer. Thus, the application of such chemical helps in improving the water status in the plants and soil.

Limitations of anti-transpirants

- May reduce the rate of photosynthesis
- May increase the leaf temperature by reducing evaporative cooling
- Interaction of climatic factors with antitranspirants reduces their effectiveness for longer duration
- Sometimes marginal cost is more than marginal returns.
- May produce toxic effects on leaves.

Demonstration of transpiration phenomenon

Loss of water in the form of water vapour from the aerial parts of plant is known as transpiration. The water vapour can be seen in the form of water droplets if a transpiring plant covered thoroughly

Material required: A well watered potted plant, belljar, rubber sheet, grease, or vaseline

Procedure

Take a small well watered plant. Cover the external soil surface of pot and its soil thoroughly with polythene bag to prevent direct evaporation of water. Invert a dry bell jar over it. Seal the edges of bell jar with vaseline and place it in sunlight for few hours. After some time the bell jar becomes misty and its inner walls contains drops of water that may flow down the sides of bell jar. Weigh the pot. Weight of the pot is less as compared to its initial weight. This reduced weight is due loss of water in transpiration.

Precaution

- The bell jar should be air tight at rim base by putting grease or Vaseline.
- The pot should be well watered before starting the experiment
- The bell jar should be made up of glass to penetrate light for photosynthesis

The pot surface and soil must be covered carefully to avoid evaporation losses.

Exercise No. 6 : Soil Moisture determination

Soil is a heterogeneous mass and consists of three phases, viz., solid, liquid and gaseous phase which serves as a water reservoir or bank. Determination of soil moisture content is needed to help in the estimation of

- Available water in the root zone,

- Scheduling of irrigation,

- Soil water potential and

- Changes in physical and chemical properties of soil due to changes in water content.

Broadly, the methods used for determination of the soil water content are classified as Direct methods and Indirect methods.

i) **Direct methods:** In these methods the techniques involve either the direct measurement of water present in a soil sample or the loss of water from a soil sample. These methods involve the removal of water from the soil samples by evaporation, leaching or chemical reaction, where soil and water are separated, and the amount of water removed is measured or the change in weight of the soil due to moisture loss is estimated. Gravimetric methods fall in this category.

ii) **Indirect methods:** These methods involve measurement of soil properties or determination of properties of some object placed in the soil, which are affected by the soil water content. Indirect methods are preferred over direct methods because they are non-destructive, less time consuming and permit more frequent observations, however, these methods require an initial calibration.

The classification of various soil moisture determination methods is as below:

   i.   Direct method i.e Gravimetric method

   ii.   Indirect methods i.e. Neutron moisture meter

Direct methods of soil moisture determination (Gravimetric method)

A. Weight basis

Materials: Sampling screw auger, moisture box, weighing balance and drying oven.

Procedure

   i.   Take a composite sample of soil about 50 g in a moisture box and cover it immediately with its lid.

298 Rainfed Agriculture

ii. Cover the box with a cloth in the field to avoid heating due to insulation. Carry the samples to the laboratory.

iii. Tipping off some soil near the lid, weigh the sample on a balance correct to two decimal places in gram. If field balance is available, the moist soil sample can be weighed just after sampling in the field (wet weight) .

iv. Dry the sample in an oven to a constant weight at 105°C. This takes about 8 hours. Record the weight of the dried sample (dry weight).

v. Weight the empty moisture box.

vi. Calculate the moisture percentage on weight basis by the formula:

$$= \frac{\text{Mass of water in soil sample}}{\text{Mass of dry soil sample sample}} \times 100$$

B. Volume basis

Materials: Sampling tube or a core sampler, moisture box, weighing balance and drying oven.

Procedure

i. Take the soil sample with the help of core sampler of known volume (V).

ii. Follow the steps for determination of soil moisture as in weight basis.

iii. Calculate the moisture percentage by the relationship,

$$= \frac{\text{Mass of water in soil sample}}{\text{Mass of dry soil sample sample}} \times BD \times 100$$

Where BD is the bulk density.

C. Depth basis

Materials: Sampling tube or a core sampler, moisture box, weighing balance and drying oven.

Procedure

i. Take the soil sample with the help of core sampler of known volume (V) up to desired depth

ii. Follow the steps for determination of soil moisture as in weight and volumetric basis

iii. The moisture content on depth basis can be calculated by the formula.

$$= \frac{\text{Mass of water in soil sample}}{\text{Mass of dry soil sample}} \times BD \times D \times 100,$$

Where BD is the bulk density and D is the depth of soil in cm..

Observations

i. Weight of the moisture box with wet sample = …………..g

ii. Weight of the moisture box after drying sample = …………..g

iii. Weight of the empty moisture box = …………..g

iv. Bulk density soil= …………..g/cc

v. Depth of soil profile=…………..cm

300   Rainfed Agriculture

Exercise No. 7 : Soil moisture determination by neutron moisture meter

This is an indirect method of soil moisture determination which uses the interaction of neutron with water molecules as a basis of monitoring soil water content in situ.

Materials: Neutron moisture meter, aluminum access tubes, balance, oven, aluminum boxes and screw auger/tube auger.

Procedure

i. Install the aluminum access tube of specific diameter with closed bottom, up to the crop rooting depth. For this, a hole of diameter equal to outer diameter of the access tube is made in the field. The access tube is inserted in it, keeping-20-30 cm above the ground. Generally a 2-meter length access tube is installed. The gap between outer circumference of the tube and inner wall of the hole is filled with soil slurry in order to ensure proper contact of the access tube with the field soil. Cover the opening of the tube with a rubber stopper.

ii. Before taking actual observation for the soil moisture, take standard count (Ns)by placing the neutron moisture meter on the top of the open access tube,keeping the instrument in ON position and timer at 4 minutes/calibration mark.

iii. Lower down the probe into the access tube up to the desired soil depth. Put the instrument on measurement mode and the timer at the desired option of 4minutes or 1 minute or 30 seconds. After the desired time record the measured count (Nm). Continue the same process to measure the count at lower depths.

iv. Find out soil moisture against the count ratio (Nm/Ns) from the calibration curve prepared for this soil by measuring various count ratios and corresponding water content values at different soil depths. Obtain directly the soil moisture from the instrument in case there is built in microprocessor in the unit.

Calibration

Initially, standard counts are taken with the probe inside the field. The shield acts as a standard having moisture 50 per cent by weight. Counts are generally taken for 1 minute intervals but for precision, they should be taken for greater intervals. Observations are taken successively at 30 cm intervals after lowering the probe in the access tube. Count ratio is then computed by dividing the observed counts (Nm) by the standard counts (Ns). The Calibration curve is obtained by taking series of counts and soil samples for gravimetric estimation simultaneously at different depths.

Merits

i. It is very quick, non-destructive method.

ii. It measures the soil water content on volume basis, in the whole range from saturation to wilting point.

iii. It is not influenced by hysteresis or salts present in soil.

Limitations

i. It is an expensive technique.

ii. Surface measurements are normally unreliable.

iii. It gives over estimation of soil moisture content in the presence of high amount of organic matter, cadmium, boron *etc.* and

iv. Requires careful handling.

302 Rainfed Agriculture

Exercise No. 8 : Determination of moisture deficit in soil profile

Since, the net irrigation requirement or soil moisture deficit is the difference between the soil moisture content before irrigation and that at field capacity, therefore, it can be determined using the following formula:

$$= \sum_{i=1}^{n} \frac{(Mfc - Mbi) \times BD \times Di}{100}$$

Where, Mbi = Moisture content before irrigation

Mfc = Moisture content at field capacity

B.D. = Bulk density of soil

Di = Depth of i_{th} layer of soil

The plant extracts moisture from the entire root zone in the soil. Hence, it is necessary to determine the moisture deficit before irrigation in the whole root zone.

Materials: Core sampler, soil moisture boxes, weighing balance, black polythene sheet or straw mulch, pressure plate apparatus, hot air oven.

Procedure

i. Select a representative spot in the field for soil moisture determination before irrigation

ii. Draw soil samples with core sampler from soil depths *viz.*, 0-15, 15-30, 30-60, 60-90 and 90-120 cm.

iii. Determine the soil moisture content and bulk density of the soil sample from each layer.

iv. Bund an area of about 2 x 2 m area, another representative spot in the same field and saturate it with water upto the effective root zone depth.

v. Cover the area with black polythene sheet or straw mulch to prevent evaporation of water from the soil surface.

vi. Draw soil samples from the same soil depths as above and determine the field capacity of each layer by field or pressure plate method.

vii. Calculate the soil moisture deficit.

Precautions

i. Select the spot well inside the field away from the borders.

ii. Draw the samples from the centre of the crop rows, if sampling is done in a cropped field.

iii. Use pressure plate method for quick determination of field capacity.

304   Rainfed Agriculture

Exercise No. 9 : Quantum of water required by plants and on farm soil and water conservation measures

Water requirement of a crop is the quantity of water needed for normal growth, development and yield and may be supplied by precipitation or by irrigation or by both. Water is needed mainly to meet the demands of evaporation (E), transpiration (T) and metabolic needs of the plants.

The water requirement of any crop is dependent upon

- Crop factors like variety, growth stage, duration, plant population and growing season.
- Soil factors like texture, structure, depth, and topography.
- Climatic factors like temperature, relative humidity and wind velocity.
- Crop management practices like tillage, fertilization, weeding *etc.*

Principles of water management

1. Where precipitation is less than crop requirement, here the strategy includes land treatment to increase run off into cropped area.
2. Where precipitation is equal to crop requirement, here the strategy is to conserve moisture within soil profile.
3. Where precipitation is more than crop requirement, in this case strategy is to reduce the rainfall erosion.

Water loss from the field

- Absorption by soil and plants.
- Storage as surface water body.
- Evapotranspiration.
- Ground water recharge.
- Ground water overflow.

On farm soil and water conservation measures

Tillage: Tillage is one of the major parts of land preparation, provides suitable seed bed for plant growth and helps to control weeds. It also influences intake rate of water and reduce soil loss by creating obstruction to surface flow.

Mulching: It is the covering of soil surface which helps to conserve soil moisture and maintain the moisture regime in the root zone area by reducing the evaporation loss Reduced evaporation potential under mulch reduces the loss of moisture from sub soil as well as from under ground water by inhibiting the capillary rise. Mulching by organic material/residue is found to be the best. Excellent results can be obtained by covering 70-75 % of the cropped area.

Contouring: Depending upon the suitability there are many types of contouring methods like contour cultivation, contour bund, contour ridge, contour trench, contour strip cropping, terracing etc. which are effective in dry farming.

Trenching or pitting: In this method the trenches or pits are dug out across the slope and along the contour, suitable to the vegetation/crop present in the area. These structure store runoff water as well eroded soil by reducing the runoff velocity and providing more time for its percolation.

Terracing: Terraces are the large (in width) plain surfaced earth embankments constructed across the slope to intercept surface runoff and convey it to suitable outlet at a non erosive velocity and to shorten slop length.

Waterways: The safe passage of excess water collected from surface runoff is generally provided through channels called waterways. They are diversion channel, terrace channel and grass water ways or main waterways.

Farm ponds: This excavated or dug out structures are usually constructed in the lower most part or natural depression of the farm and in geometrical shapes like sqare, rectangular or circular.

Check dams: It is a small permanent type of obstruction that checks the flow of water running through any water course and intercepts the eroded soil carried by it. These structures are placed across the nala or gully with necessary passage for excess water to flow.

306 Rainfed Agriculture

Exercise No. 10 : Critical growth stages to water stress for major field crops

Crop	Crop duration (days)	Critical stages for irrigation
Wheat (dwarf)	130-140	Crowing, flowering, jointing, milk and dough
Wheat (tall)	140-150	Tillering, heading and dough
Barley	125-135	Tillering, booting
Rice (dwarf)	130-140	Flowering, grain filling, tillering
Rice (Basmati)	140-150	Flowering, grain development
Summer maize	90-110	Tasseling and silking, and grain development
Winter maize	150-180	Tasseling, grain development
Pearlmillet	75-85	Heading
Sorghum	90-110	Flowering, seeding, grain filling
Cotton	150-180	Flowering and boll development
Chickpea	130-140	Preflowering and pod development
Pigeonpea	150-170	Flowering initiation, pod development
Moongbean	65-75	Flowering
Clusterbean	130-145	Branching and flowering
Mustard	140-150	Pre-flowering and pod development
Groundnut	110-135	Pegging, pod-setting and pod filling
Sunflower	100-120	Flowering, seed filling, late vegetative
Sugarcane	280-330	Tillering and grand growth

Water requirement (WR) of major dryland crops and fruit trees

Summer crops	WR(mm)	Winter crops	WR(mm)	Fruit crops	WR(mm)
Pearlmillet	250	Rocket plant	150	Gonda	560
Mungbean	225	Mustard	200	Ber	660
Clusterbean	275	Chickpea	250	Aonla	660
Cowpea	240	Lentil	285	Clusapple	860
Pigeonpea	450	Peas	300	Guava	1160
Maize	400	Barley	300	Pomegranate	1160
Sorghum	400	Wheat	350	Mango	1260

Practical Excercises 307

Exercise No.11 : Study of cropping systems for dryland areas

Cropping system is an important component of a farming system. It represents cropping patterns used on a farm and their interaction with farm resources, other farm and available technology which determine their make up. Cropping pattern means the proportion of area under various crops at a point of time in a unit area. It indicates the yearly sequence and spatial arrangement of crops and fallow in an area. Crop sequence and crop rotation are generally used synonymously. Crop rotation refers to recurrent succession of crops on the same piece of land either in a year or over a longer period of time. Component crops are so chosen that soil health is not impaired.

Types of cropping systems

Depending on the resources and technology available, different types of cropping systems are adopted on farms.

Monocropping

Monocropping or monoculture refers to growing of only one crop on a piece of land year after year. Under rainfed condition ground nut or cotton or sorghum are grown year after year due to limitation of rainfall. Mono cropping is practiced under normal and below normal rainfall conditions by adopting improved methods of moisture conservation practices and short and medium duration varieties of the crops, either in kharif or rabi season.

Monocropping in normal and below normal rainfall years.

Kharif crops – fallow

Fallow- rabi crops (on conserved moisture)

Multiple cropping

Growing two or more crops on the same piece of land in one calendar year is known as multiple cropping. It is the intensification of cropping in time and space dimension i.e. more number of crops within year and more number of crops on the same piece of land at any given period. It includes double cropping, intercropping , mixed cropping and sequence cropping.

Use of seeds of drought resistant, early, quick growing and high yielding varieties of the crops with improved agronomic management and cropping system depending upon the rainfall pattern bring about definite improvement in dryland crop yields. Pearlmillet, mungbean, black gram, cowpea and clusterbean during rainy season and chickpea, mustard and taramira during rabi season were identified to be the promising crops for the dryland areas of the region.

308 Rainfed Agriculture

Double cropping in above normal rainfall years

Pearlmillet- chickpea

Mungbean-mustard

Cowpea-mustard

Pearlmillet+ cowpea- mustard/chickpea

Intercropping

Pearlmillet + mungbean

Pearlmillet + cluster bean

For post rainy season crops grown on conserved soil moisture, it is the available soil moisture in the soil profile at sowing time that dictates the choice of crops.

Mustard (RH-30, Varuna, RH-819 and Lakshmi) and gram (H-208 and C-235) were identified most suitable crops for planting on well conserved soil moisture.

i) Under limited moisture availability i.e. less than 300 mm m^{-1} the response of grain yield of wheat, barley, gram, mustard and taramira to water is linear. In case of taramira this response is negative as the crop lodges due to luxuriant growth when more moisture is available.

ii) For each mm of increased available water to the crops, wheat, barley, gram and mustard yield increases but in taramira with the increasing availability of water beyond 150 mm per meter yield decreases.

iii) Wheat crop needs more than 250 mm m^{-1} water in the soil profile for obtaining satisfactory yield. The requirement of barley is about 200 mm m^{-1} and that of gram is 150-200 mm m^{-1}. For soil moisture between 125-175 mm m^{-1} mustard can be raised successfully and for low moisture (<125 mm m^{-1}) in soil profile, taramira is the only crop that can be successfully grown.

Exercise No. 12 : Description of dryland areas of Haryana and production/productivity of important dryland crops

Table 1: District wise gross irrigated area in Haryana

Sr. No.	District	Gross irrigated area (ha)
1.	Ambala	186
2.	Panchkula	23
3.	Yamuna Nagar	192
4.	Kurukshetra	275
5.	Kaithal	381
6.	Karnal	387
7.	Panipat	188
8.	Sonipat	289
9.	Rohtak	191
10.	Jhajjar	194
11.	Faridabad	206
12.	Gurgaon	104
13.	Mewat	150
14.	Rewari	163
15.	Mahendragarh	147
16.	Bhiwani	423
17.	Jind	441
18.	Hisar	541
19.	Fatehabad	411
20.	Sirsa	661
Total		5553

Table 2 : Year wise gross irrigated area in India

Year	Gross irrigated area (m ha)
1998-99	78670
1999-00	79216
2000-01	76187
2001-02	78420
2002-03	73411
2003-04	78147
2004-05	81181
2005-06	83939
2006-07	86504
2007-08	87259

310 Rainfed Agriculture

Table 3 : District wise gross unirrigated area in Haryana

Sr. No.	District	Gross unirrigated area ('000 ha)
1.	Ambala	20
2.	Panchkula	24
3.	Yamuna Nagar	22
4.	Kurukshetra	4
5.	Kaithal	4
6.	Karnal	8
7.	Panipat	3
8.	Sonipat	12
9.	Rohtak	44
10.	Jhajjar	94
11.	Faridabad	-
12.	Gurgaon	10
13.	Mewat	22
14.	Rewari	-
15.	Mahendragarh	34
16.	Bhiwani	106
17.	Jind	336
18.	Hisar	29
19.	Fatehabad	107
20.	Sirsa	13
Total		893

Table. 4 : State wise per cent unirrigated and irrigated area in India

Sr. No.	State	Irrigated area (%)	Unirrigated area (%)
1.	Andhra Pradesh	43.9	56.1
2.	Arunachal Pradesh	24.9	75.1
3.	Assam	8.1	91.9
4.	Bihar	56.9	43.1
5.	Jharkhand	9.3	90.7
6.	Goa	17.5	82.5
7.	Gujarat	34.4	65.6
8.	Haryana	84.1	15.9
9.	Himachal Pradesh	19.2	80.8
10.	J&K	41.6	58.4
11.	Karnataka	29.2	70.8

12.	Kerala	18.7	81.3
13.	M.P.	43.2	56.8
14.	Chattishgarh	27.1	72.9
15.	Maharashtra	16.9	83.1
16.	Manipur	22.8	77.2
17.	Maghaleya	31.5	68.5
18.	Mizoram	17.4	82.6
19.	Nagaland	20.2	79.8
20.	Orrisa	32.2	67.8
21.	Punjab	94.9	5.1
22.	Rajasthan	38.7	61.3
23.	Sikkim	8.0	92.0
24.	Tamilnadu	56.4	43.6
25.	Tripura	21.8	78.2
26.	U.P.	78.6	21.4
27.	Uttrakhand	44.7	55.3
28.	West Bengal	59.2	40.8

Table. 5 : Extent of area under different crops in unirrigated or rainfed condition

Crop	Unirrigated area (%)
Total pulses	83.8
Total cereals	45.70
Total oilseeds	72.90
Total food grains	53.20

Table 6 : Agro-ecological conditions in different rainfall zones in India

Annual rainfall(mm)	Soil type	Mean annual temperature	Length of growing season (days)	Dominant crops
< 400	Sandy soil	0-10	30-90	Pearlmillet, short duration pulses
400-1000	Sandy, red,black	20-30	80-200	Pearlmillet, sorghum, maize, oilseed, pulses, cotton
1000-1800	Alluvial, laterite	20-30	200-300	Maize, paddy, wheat, barley, mustard, sorghum,
>1800	Submountain	15-20	>300	Paddy, plantation crop such as tobacco, tea

312 Rainfed Agriculture

Table 7 : Important dryland crops growing states in decreasing order

Crop	Rank-I	Rank-II	Rank-III	Rank-IV	Rank-V
Pearlmillet	Rajasthan	Maharastra	Gujrat	U.P	Haryana
Sorghum	Maharastra	Karnataka	M.P.	A.P.	Rajasthan
Chickpea	M.P.	U.P	Rajasthan	Maharastra	Haryana
Mustard	Rajasthan	U.P	M.P.	W.B.	Haryana
Cotton	Maharastra	Gujrat	Punjab	A.P.	Karnataka

Table 8 : Rainfed area intensity in various districts of Haryana

High (>40%)	Moderate (20-40%)	Low (10-20%)	Negligible (<10%)
Gurgaon (64.4)	M.Garh (31.8)	Ambala (19.7)	Rohtak (7.3)
Panchkula (54.5)	Jhajjar (31.6)	Y.Nagar (16.1)	Sonipat (5.1)
Bhiwani (45.9)	Hisar (25.2)	Jind (15.2)	Panipat (3.0)
	Rewari (25.0)	Faridabad (13.9)	Karnal (0.5)
	Sirsa (21.4)	Fatehabad (11.3)	Kaithal (0.0)
			Kurushetra (0.0)

Table 9 : Area coverage of various dryland crops in Haryana

Crops	Total area (Mha)	Dryland coverage (%)
Pearlmillet	0.67	85
Gram	0.46	74
Mustard	0.49	37
Clusterbean	0.12	72
Sorghum	0.14	53
Maize	0.14	18
Barley	0.03	77

Table 10 : Production and Productivity of important dryland crops in Haryana

Year	Production ('000 tonnes)			Productivity (kg/ha)		
	Pearlmillet	Mustard	Gram	Pearlmillet	Mustard	Gram
1970-71	826	89	789	939	678	742
1975-76	608	65	907	605	471	820
1980-81	474	178	455	544	634	629
1985-86	315	276	625	487	798	821
1990-91	526	634	469	864	1338	722
1995-96	409	729	381	711	1198	1010
2000-2001	656	560	80	1079	1369	640
2005-2006	706	793	72	1117	1117	554
2008-2009	1087	895	129	1773	1722	1040

Practical Excercises 313

Exercise No. 13: Seed treatment, seed germination and crop establishment in relation to soil moisture contents

Seed germination means the resumption of growth by embryo and development of a young seedling from the seed. Germination is an activation of dormant embryo to give rise to radical (root development) and plumule (stem development). Germination is the awakening of the dormant embryo. The process by which the dormant embryo wakes up and begins to grow is known as germination. Seed emergence means actually coming above and out of the soil surface by the seedling.

Changes during germination

1) Swelling of seed due to imbibitions of water by osmosis.

2) Initiation of physiological activities such as respiration and secretion of enzyme.

3) Digestion of stored food by enzymes.

4) Translocation and assimilation of soluble food.

When seed is placed in soil gets favorable conditions, radical grows vigorously and comes out through microphyle and fixes seed in the soil. Then either hypo or epicotyls begins to grow.

Factors affecting the germination

External factors

1. Moisture: It enables the resumption of physiological activities, swelling of seed due to absorption of moisture and causes bursting of seed coat and softening the tissue due to which embryo awakes and resumes its growth.

2. Temperature: A suitable temperature is necessary for proper germination. Germination does not take place beyond certain minimum and maximum temperature i.e. 0°C & above 50°C. Optimum temperature range for satisfactory germination of seed is 25 to 30°C.

3. Oxygen: It is essential during germination for respiration and other physiological activities which are vigorous during the process.

4. Light: It is not considered as essential for germination and it takes place without light. The seedlings grow more vigorously during darkness rather in light. However, for survival of germinating seedling, light is quite essential.

5. Substratum: It is the medium used for germinating seeds. In the laboratory,

Rainfed Agriculture

it may be absorbent paper (blotting paper, towel or tissue paper), soil and sand. Substratum absorbs water and supplies to the germinating seeds. It should be free from toxic substances and should not act as medium for growth of micro-organisms.

Seed dormancy

Failure of fully developed and mature viable seed to germinate under favorable conditions of moisture and temperature is called resting stage or dormancy and the seed is said to be dormant.

Kinds of dormancy in seeds

1. Primary dormancy: The seeds which are capable of germination just after ripening even by providing all the favorable conditions are said to have primary dormancy. E.g.: Potato.

2. Secondary dormancy: Some seeds are capable of germination under favorable conditions just after ripening but when these seeds are stored under unfavorable conditions even for few days, they become incapable of germination.

3. Special type of dormancy: Sometimes seeds germinate but the growth of the sprouts is found to be restricted because of a very poor development of roots and coleoptiles.

Seed treatment

Seed treatment refers to the application of fungicide, insecticide, or a combination of both, to seeds so as to disinfect and disinfest them seed-borne or soil-borne pathogenic organism and storage insects. It also refers to the subjecting of seeds to solar energy exposure, immersion in conditioned water, *etc*. The seed treatment is done to achieve the following benefits.

1. Benefits of seed treatment

1. Prevention of spread of plant diseases.

The disease from treatment standpoint may be conveniently grouped under two types:

a) Systemic disease

That infect the seed during the harvest or storage period resulting in infection of seed, *e.g.* Bunt or stinking smut of wheat, *Helminthosporium* blight of barley, loose and covered smut of oats; head and kernel smuts of rye, smuts of millet. Appropriate seed treatment is significantly effective in controlling these diseases. That infects seed during the flowering stage to become established within the

Practical Excercises 315

seed and from there within the resulting plant. Such diseases include loose smuts of wheat. Treatment with systemic fungicides, E.g Vitavax has been found effective.

c) Non-systemic disease

Diseases that infect seed during the harvest or storage period. Such diseases includes *Helminthosporium* blight, blotches or blight of barley, oats, rice , rye , sorghum, wheat and *Fusarium*. These diseases can be effectively controlled by appropriate seed treatment.

2. Reduces soil and seed borne diseases

Protects seed from seed rot and seedling blights. Seed treatment, by its protective coating around the seed, acts as a barrier once the seed is planted to ward off attack by both seed-borne and soil-borne organisms. These organisms affect, all crop seeds and the degree of attack depends upon a number of factors of particular importance are the organisms. *Pythium spp, Rhizoctonia* and *Sclerotium* that are present in all soils. They may rot the seed before germination gets well started, or they may kill the seedling before it emerges, or so affect it that it dies after emergence or supervives only as a weakened plant. The responses to protective treatment varies with the kind of crop seed, the vigour of the particular seed, the amount of mechanical injury to seeds, conditions of seed surface and adversity of planting conditions. The fungicide treatment compensates by protecting these cracks and abrasions from entrance of fungi.

3. Improves germination

Seed treatment often improves the standard of germination through the control of seed surface flora, though normally not considered pathogenic; this may infect the seed following moist harvesting and storage conditions. In the germination test it may smother the seed before it has a chance to germinate.

4. Provides protection from storage insects

The protection of seed from insect damage during storage is of increasing importance with the trend towards processing, treating and unit packaging of seeds at harvest time. For complete protection it is necessary to treat seed with insecticide also. Insecticides are more needed in warm storage than cool storage.

5. Controlling soil insects

This can be done through combination treatment – the process of addition of an insecticide with fungicide for the added protection of the seed and seedling against certain soil insects, such as wire worm and the seed corn maggot. In contrast to storage insect protection, it is a means of giving limited protection to the seed and seedling until it becomes resistant to attack or can survive limited attack. It is not a means of disinfecting the soil.

Following factors are to be taken into consideration for seed germination

1) **Dormancy period:** A seed requires to undergo a period of dormancy after its formation. During this period the seed 'rests' and may refuse to germinate properly. This period varies from seed to seed. Seeds of water melon or red gourd do not have a dormany period and may germinate easily immediately after their formation in the fruit. Others require dormancy period ranging from a few weeks to a few months.

2) **Viability of a seed:** It is exactly like expiry date/period of a commercial product. It must not be confused with the dormancy or 'resting period' after the resting period a seed becomes ready to germinate. But this readiness is not forever. It literally expires after certain period. After that seed will not germinate. This 'ready to germinate' period is the viability period of a seed. It may vary from a few years to many centuries. Seeds of Lotus are reported to have germinated even after eight hundred years! In that case their viability period is at least that much if not more.

External factors

3) **Soil moisture or water:** This is required to trigger the mechanism of germination. In the absence of moisture the seeds cannot germinate; but when it is available it is imbibed by the seed coat and the enzymes in side become active and functional. The amount of water does not matter in the initial stages; but later on it becomes critical.

4) **Soil Texture:** Does not matter during germination. Seeds can even germinate without soil on the piece of a moist blotting paper.

5) **Soil pH:** Should too acidic. pH 5 to 8 is alright.

6) **Soil temparature:** It should be on the warmer side. Lesser the temp. lesser the rate of germination.

7) **Light:** It does not affect the process of germination; but some seeds do germinate better in the absence of light. They like darkness for germination it seems.

Exercise No. 14: Study of moisture stress effects and recovery behavior of important crops

Introduction

Biosphere's continued exposure to abiotic stresses, for example, drought, salinity, extreme temperatures, chemical toxicity, oxidative stress, etc., cause imbalances in the natural status of the environment. Each year, stresses on arable plants in different parts of the world disrupt agriculture and food supply with the final consequence - famine. Factors controlling stress conditions alter the normal equilibrium, and lead to a series of morphological, physiological, biochemical and molecular changes in plants, which adversely affect their growth and productivity. The average yields from the major crop plants may reduce by more than 50% owing to stresses. However, plants also have developed innate adaptations to stress conditions with an array of biochemical and physiological interventions that involves the function of many stress-associated genes. In this chapter, we aim at the stresses related to water and the expression 'drought' which is derived from the agricultural context, is used as equal to water stress throughout the article. Water, comprising 80-90% of the biomass of non-woody plants, is the central molecule in all physiological processes of plants by being the major medium for transporting metabolites and nutrients. Drought is a situation that lowers plant water potential and turgor to the extent that plants face difficulties in executing normal physiological functions. However, a few groups of animals and a wide variety of plants are known for their tolerance to desiccation during the adult stages of their life cycle.

Water stress – Why and how?

Plants experience water stress either when the water supply to their roots becomes limiting or when the transpiration rate becomes intense. Water stress is primarily caused by the water deficit, *i.e.* drought or high soil salinity. In case of high soil salinity and also in other conditions like flooding and low soil temperature, water exists in soil solution but plants cannot uptake it – a situation commonly known as 'physiological drought'. Drought occurs in many parts of the world every year, frequently experienced in the field grown plants under arid and semi-arid climates. Regions with adequate but non-uniform precipitation also experience water limiting environments. Since the dawn of agriculture, mild to severe drought has been one of the major production limiting factors. Consequently, the ability of plants to withstand such stress is of immense economic importance. The general effects of drought on plant growth are fairly well known. However, the primary effect of water deficit at the biochemical and molecular levels are not considerably understood yet and such understanding is crucial. All plants have tolerance to water stress, but the extent varies from species to species. Knowledge of the biochemical and molecular responses to drought is essential for a holistic perception of plant resistance mechanisms to water limited conditions in higher plants.

Water stress effect on plants

Water stress causes leaves to wilt, dry and fall off. Lack of water is the single most limiting factor in the regulation of photosynthesis. Photosynthesis is the process by which plants convert light energy, carbon dioxide and water to plant sugars. In the absence of water, no amount of light or carbon dioxide will be of use to the plant. Leaves wilt, photosynthesis mechanisms shut down and growth slows to a halt.

1. Wilting

Plant cells contain large vessels known as vacuoles. These structures serve a number of functions, including water storage. Well-hydrated vacuoles push against the thick cell walls making the cell rigid, or turgid. When all cells are turgid, the plant itself is firm and crisp. When a plant loses water, the vacuole contracts and cannot maintain this pressure. The cells become flaccid and the plant wilts.

2. Closed stomata

When plants are well-hydrated, up to 95 percent of water absorbed through the roots returns to the atmosphere by transpiration. Transpiration is the release of water vapor and oxygen gas through tiny holes in the leaves known as stomata. When moisture is limited, the stomata close to slow transpiration and conserve water. Photosynthesis cannot occur when stomata are closed, and growth stops.

3. Increased susceptibility to photoinhibition

Overexposure to sunlight can reduce photosynthetic efficiency. Botanists refer to this as photoinhibition. Experts once believed that closed stomata alone accounted for reduced photosynthesis under drought conditions, but a study published in the "Australian Journal of Plant Physiology" indicated that drought

Practical Excercises 319

alters physiological functions as well. Water stress increases the proportion of photosynthetic elements most susceptible to light damage, making the entire plant more vulnerable.

4. Loss of leaves

When water is scarce, aging foliage becomes a liability to the plant. Deciduous plants may undergo early senescence, in which leaves go through the natural shedding process ahead of schedule. Other plants may simply shed wilted leaves. Symptoms of drought-induced defoliation include curling, rolling, folding and eventual shedding of the leaves.

5. Root damage

The root hairs primarily responsible for water uptake are also the most susceptible to damage. The University of Georgia School of Forest Resources explains that during the early stages of water stress, root growth increases to access water deeper in the soil. As drought continues, this extra surface area leads to loss of water. Plants respond by converting root hairs to woody cork, making them unsuitable for water transport.

6. Vulnerability to pest damage

Plants that are weakened from water stress are more susceptible to damage from pest species and plant disease. The University of Georgia asserts that stressed trees have fewer resources available for defence against invading insects and pathogens or for recovery from damage.

7. Slowed growth

Root damage, pest invasion and reduced photosynthesis slow plant growth. Shoot elongation and formation, bud formation and leaf development cannot take place with inadequate resources. This can be especially critical for tree species that have limited windows of growth during the year. For these species, a brief summer drought can impact the entire year's growth.

How can plant water stress be managed?

Crop selection can be a key component when dealing with or anticipating moisture stress. Generalizations about plant groups and how they behave under moisture stress can be used to guide decisions about crop selection for drought and saline conditions.

- **Determinate crops:** Resistant to moisture stress during vegetative stages, determinate crops are grown for harvest of mature seed and include small grains, cereal crops, peas, beans, and oil seed crops. Determinate crops show a linear relationship between water stress and seed production. These crops are most sensitive to stress during seed formation including heading, flowering, and pollination. Each has a minimum threshold growth and water requirement for seed production. This process can be interrupted by stress and generally can't be recovered with removal of the stress.

- **Indeterminate crops:** Indeterminate crops include tubers and root crops such as potatoes, carrots, and sugar beets. These crops are relatively insensitive to moisture stress in short intervals (4-5 days) throughout the growing season and have no specific critical periods. If and indeterminate crop is subject to moisture stress, quality will be affected rather than yield. Harvestable yield increases as water use increases. Indeterminate crops are more directly related to climatic demand and cumulative water use during the season than to stress during any particular growth stage.

- **Forages:** Forage crops are grown for hay, pasture, and biomass production. In comparison to determinate and indeterminate crops, perennial forages are impacted least by moisture stress. Perennials usually have deep well established roots systems. Forage yields are typically in response to climatic conditions. Forages that have undergone moisture stress will have lower yields than those that have not. Annual forages are an effective way to take advantage of early season moisture and cooler temperatures. In general, as water stress is increased, forage nutritional value is increased, yet overall yield and harvestable protein is decreased.

Table 1: Crop selection for water stress management

Crop type	Water stress limitations	Management tips
Determinate crops	Resistant to water stress during vegetative stages.	Avoid water stress during reproductive stages.
Indeterminate crops	No specific critical periods.	Sugarbeets are more stress tolerant than potatoes, carrots, and onions.
Forages	Perennial forages are least affected by moisture stress in the long run.	Concentrate irrigation efforts early in the season to maximize production.

Exercise No.15 : Estimation of moisture index and aridity index

Moisture index: A measure of the water balance of an area in terms of gains from precipitation (P) and losses from potential evapotranspiration (PE). The moisture index (MI) is calculated thus:

$$MI = 100(P - PE)\, PE$$

An **Aridity index** (AI) is a numerical indicator of the degree of dryness of the climate at a given location. A number of aridity indices have been proposed ,these indicators serve to identify, locate or delimit regions that suffer from a deficit of available water, a condition that can severely affect the effective use of the land for such activities as agriculture or stock-farming.

Indices

In 20th century, Wladimir Köppen and Rudolf Geiger developed the concept of a climate classification where arid regions were defined as those places where the annual rainfall accumulation (in centimetres) is less than $R\, /\, 2$, where:

- $R = 2 \times T$ If rainfall occurs mainly in the cold season,
- $R = 2 \times T + 14$ If rainfall is evenly distributed throughout the year, and
- $R = 2 \times T + 28$ If rainfall occurs mainly in the hot season.

where T is the mean annual temperature in Celsius.

In 1948, C. W. Thornthwaite proposed an AI defined as:

$$AI_T = 100 \times \frac{d}{n}$$

Where the water deficiency d is calculated as the sum of the monthly differences between precipitation and potential evapotranspiration for those months when the normal precipitation is less than the normal evapotranspiration; and where n stands for the sum of monthly values of potential evapotranspiration for the deficient months (after Huschke, 1959). This AI was later used by Meigs (1961) to delineate the arid zones of the world in the context of the UNESCO Arid Zone Research programme.

In the preparations leading to the UN Conference on Desertification (UNCOD), the United Nations Environment Programme (UNEP) issued a dryness map based on a different aridity index, proposed originally by Mikhail Ivanovich Budyko (1958) and defined as follows:

$$AI_B = 100 \times \frac{R}{LP}$$

Where R is the mean annual net radiation (also known as the net radiation balance), P is the mean annual precipitation, and L is the latent heat of vaporization for water. Note that this index is dimensionless and that the variables R, L and P can be expressed in any system of units that is self-consistent.

More recently, the UNEP has adopted yet another index of aridity, defined as:

$$AI_U = \frac{P}{PET}$$

where PET is the potential evapotranspiration and P is the average annual precipitation. Here also, PET and P must be expressed in the same units, e.g., in milimetres. In this latter case, the boundaries that define various degrees of aridity and the approximate areas involved are as follows :

Classification	Aridity index	Global land area
Hyperarid	AI < 0.05	7.5%
Arid	0.05 < AI < 0.20	12.1%
Semi-arid	0.20 < AI < 0.50	17.7%
Dry subhumid	0.50 < AI < 0.65	9.9%

Exercise No. 16 : Collection and interpretation of data for water balance equations

Water is essentially required for different life forms such as plants, animals, birds etc. for cell building and other purposes. The main source for the water is ocean. The water from the oceans is evaporated, clouds are formed and carried away by wind and they precipitate. The water received from precipitation is lost to the ocean back by different processes such as run- off evaporation from soil, lakes and ponds, streams, etc evapotranspiration from plants the water which is absorbed in the ground is also lost by direct or indirect way to the ocean, for example, some water which is absorbed in ground is utilized by plants and then evaporated, the ground water which is absorbed in ground is utilized by plants and then evaporated. The ground water flows to the streams and the stretch finally lost in the oceans etc. Thus, we find that there is a constant circulation of water from oceans to the air and back again to the oceans. This process has not end beginning and therefore it is termed as hydrological cycle or water cycle. The hydrological cycle can be briefed by the following equation.

$$P = ET+DST + S$$

The total amount of water present on the earth surface remains constant but undergoes continuous transformation from water vapor to liquid. This equation is also called as water balance equation. Where P is the water received by precipitation, ET is loss by evapotranspiration, DST is the gain on loss by storage in the soil and S is the surplus run-off of water, from this mathematical relation, we can find out the value of other elements.

A **water balance** equation can be used to describe the flow of water in and out of a system. A system can be one of several hydrological domains, such as a column of soil or a drainage basin. Crop water requirement (WR) is the quantity of water utilized by a crop, irrespective of its source for obtaining maximum yield in a particular area without any/minimum adverse effect on soil properties. In a cropped field water evaporates (E) from the bare soil, is transpired through plants (T) and some quantity is retained in plants body for metabolic activities (Wm), added together (E+T+Wm) known as crop consumptive use of water (CU). Water retained in crop plants at any time is a very small fraction compared to quantity lost through evaporation and transpiration, together known as evapotranspiration (ET). Both CU and ET, therefore, are used interchangeably. Besides, some losses do take place in field during application and some time water is also needed for special operations such as land preparation, transplanting, leaching of salts *etc*. The water requirement (WR) thus can be expressed as:

WR = CU + Application losses + Water needed for special operation

= Irrigation + Effective rainfall + Ground water contribution + Change in soil moisture

The former equation is termed as demand equation, whereas, the later one is the supply equation. Both equations are together termed as water balance equation.

A general water balance equation is:

$P = Q+E$

Where,

P is precipitation

Q is runoff

E is evapotranspiration

DS is the change in storage (in soil or the bedrock)

This equation uses the principles of conservation of mass in a closed system, whereby any water entering a system (via precipitation), must be transferred into either evaporation, surface runoff (eventually reaching the channel and leaving in the form of river discharge), or stored in the ground. This equation requires the system to be closed, and where is n't (for example when surface runoff contributes to a different basin), this must be taken into account.

A water balance can be used to help manage water supply and predict where there may be water shortages. It is also used in irrigation, runoff assessment (*e.g.* through the *Rain Off* model), flood control and pollution control. Further it is used in the design of subsurface drainage systems which may be *horizontal* (*i.e.* using pipes, tile drains or ditches) or *vertical* (drainage by wells). To estimate the drainage requirement, the use of a hydro geological water balance and a groundwater model may be instrumental.

Methods of computation of the main water balance components

Basic data

Records of precipitation and runoff from the network of stations are the basic data for computation of the water balance components of river basins for long-term periods. These records are published in hydrological and meteorological year-books, bulletins, *etc.* in hydrological and meteorological year-books, bulletins,

Practical Excercises   325

To compute the water balance in for individual years, seasons, or etc. months, it is necessary in addition to have data on water storage variations in the basin. These are obtained from snow surveys, observations of soil moisture, water-level fluctuations in lakes and ground-water fluctuations in wells.

To compute the water balance of small areas with special features in the water balance (mountain glacier basins, large forest areas, irrigated land, *etc.*), it is necessary in most cases to organize a special programme of observations, *e.g.* observations of glacier ablation, interception of precipitation, soil moisture, *etc.*

To compute evaporation it is desirable to have data from evaporation pans or tanks and meteorological data on temperature, humidity, wind, cloudiness, and radiation.

Exercise No. 17 : Water use efficiency (WUE)

Water use efficiency is defined as yield of marketable crop produced per unit of water used in evapotranspiration.

$WUE = Y / ET$

Where,

WUE = Water use efficiency (kg/ha/mm of water)

Y = Marketable yield (kg/ha)

ET = Evapotranspiration (mm)

If yield is proportional to ET, water use efficiency has to be constant but it is not so. Actually, Y and ET are influenced independently by crop management and environment. Yield is more influenced by crop management practices, while ET is mainly dependent on climate and soil moisture. Fertilization and other cultural practices for high yield usually increase in water use accompanying fertilization is often negligible. Crop production can be increased by judicious irrigation without markedly increasing ET. Under optimum water supply, ET is not dependent on kind of plant canopy provided the soil is adequately covered with crop.

Increasing the amount of plant canopy has therefore little or no effect on ET. Obviously, any practice that promotes plant growth and more efficient use of sunlight in photosynthesis without causing a corresponding increase in ET will increase WUE.

Factors affecting WUE

1. Nature of the plant

There are considerable between plant species to produce a unit dry matter per unit amount of water used resulting in widely varying values of WUE.

Water use efficiency of different crops

Crop	Water requirement, mm	Grain yield, kg/ha	WUE, kg/ha/mm
Rice	2000	6000	3.0
Sorghum	500	4500	9.0
Bajra	500	4000	8.0
Maize	625	5000	8.0
Groundnut	506	4680	9.2
Wheat	280	3534	12.6
Finger millet	310	4137	13.4

There is also difference in WUE between varieties of the same crop. Selection of properly adopted crop, with good rooting habit, low transpiration rates increase. WUE.

2. Climatic conditions

Weather affects both Y and ET. Manipulation of climate to any extent is possible at present. However, ET can be reduced by mulching, use of antitranspirant etc. To limited extent , but may not be economical or practical. Weed control is the most effective means of reducing ET losses and increasing the amount of water available to the crop thereby increasing WUE.

3. Soil moisture content

In adequate supply of soil moisture as well as excess moisture supply to the crop have an adverse effect on plant growth and production and therefore conductive to low WUE. For each crop combination of environment conditions, there is a narrow range of soils moisture level at which WUE is higher than with lesser or greater supply of water, proper scheduling of irrigation will increase WUE.

4. Fertilizers

Irrigation improves a greater demand for plant nutrients. Nutrient availability is highest for most of the crops when water tension is low. All available evidences indicate that under adequate irrigation suitable fertilization generally increase yield considerably, with a relatively small increase in ET and therefore, markedly improve WUE.

5. Plant population

Higher yield potential made possible by the favorable water regime provided by irrigation, the high soil fertility level resulting from heavy application of fertilizers and genetic potential of new varieties and hybrids, could be achieved only with appropriate adjustments of the population. The highest yields and WUE are possible only through optimum levels of soil moisture regime, plant population and fertilization.

Exercise No. 18 : Preparation of crop plans for different drought conditions

Crop production in dry land suffers from in-stability due to aberrant weather condition from time to time. Delayed monsoon results in non sowing of traditional kharif crops which accounts for nearly 25 to 30% of the total area under crops. So also early withdrawal of monsoon interferes sowing of rabi crop which is main constraint of crop production in the region. Similarly, breaks in monsoon also after crop production adversely which resulted in total failure of both kharif and rabi crops under dry lands. Crops and varieties selected should match the length of growing season during which they are not subjected to soil moisture stress. Climatological analysis helps to identify cultivars suitable for different regions. Feasibility for intercropping, sequential cropping and double cropping can also be known from such analysis. For regions with cropping season less than 20 weeks, single crop during kharif or rabi is recommended. Regions with more than 30 weeks and above have no problem for double cropping. In regions with 20-30 weeks cropping season, double cropping may be risky. Such areas are ideal for intercropping.

Mid - Season correction: Crop planning under aberrant weather condition in dry land.

Sr.No.	Nature of rainfall	Crops to be grown
1.	Delayed on set of monsoon	
2.	Rains during July and sowing of kharif crops by end of July or early August.	Setaria (Arjun), Red gram (No. 148) Sunflower (EC 68414), caster () Horse gram (Mans, Sinha)
3.	Rains during August and Sowing up to end of August	Red gram (No. 148), Sunflower (EC 68414) Caster (Aruna)
4.	Rains during late August & sowing up to early September.	Castor (Aruna). Jowar for fodder
5.	Good onset of monsoon	Sowing of all *kharif* crops. Common situation usually one or two dry spells are noticed
6.	If dry spell exceeds two weeks	Corrective measures a) control of plant population b) checking weed growth c) increasing interculturing serious situation in drought prone area
7.	Early withdrawal of monsoon	a) Reducing of plant population from lakh to 50 thousand as in case of rabi Jowar in 35 -1 before grand growth. b) Use of surface mulch

c) Protective irrigation 30 - 45 days drought.

d) Increase frequently of inter culturing

e) Stripping of leaves

8. Extended monsoon — It is rarely experienced

a) Sowing of grown and wheat instead of rabi sorghum.

b) Double cropping would be possible in medium deep soils.

c) Postponement of sowing of rabi crops.

Choice of crops

Traditional cropping pattern in the dry farming areas is dominated by food grains viz., millets and pulses. In a predominantly subsistence type of farming system, such dominance of food crops is natural. The choice of crops for drylands is affected by :

- Rainfall quantity and distribution

- Time of onset of rainy season

- Duration of rainy reason

- Soil characters including amount of rain water stored in the soil

- Farmer's requirements.

The major focus of research under AICRPDA has been on the identification of most efficient crops for each dry farming region. The criteria for choice of crops for dry farming regions comprise the following.

- Tolerance to drought

- Fast growth during initial period to withstand harsh environment

- Genetic potential for high yield

- Short or medium duration to escape terminal drought

- Adaptability to wide climatic variations

- Responsive to fertilizers.

For many dry farming regions of India, more suitable crops than existing ones have been identified. However, the acceptance and adoption of the practice of crop substitution by dry land farmers is poor since in most instances the new crops replace food crops.

Suggested crops in place of traditional crops

Region	Traditional crop	Yield(q/ha)	More suitable crop	Yield(q/ha)
Agra	Wheat	10.3	Mustard	20.4
Bellary	Cotton	2.0	Sorghum	26.7
Bijapur	Wheat	9.4	Safflower	18.8
Varanasi	Upland rice	28.0	Maize	33.8

Selection of suitable varieties

In most crops of dry farming regions, traditional local varieties still dominate. The preference for these local varieties is based on their pronounced drought tolerance. But they are usually longer in duration susceptible to moisture stress at maturity. They have low yield potential even under favourable rainfall. They do not respond significantly to improved management practices such as nutrient supply. The criteria now adopted for selection of crop varieties for dry lands include drought tolerance, short or medium duration, high yield potential, response to nutrient supply, high water use efficiency, moderate resistance to pest and diseases. Suitable varieties for all dry land crops have been developed in all the dry farming regions and have proved their high yield potential.

Choice of cropping system

Choice of suitable cropping system must aim at maximum and sustainable use of resources especially water and soil. Cropping systems depend on rainfall quantity, length of rainy reason and soil storage capacity. The broad guidelines in choosing a cropping system for dry lands are given below:

Potential cropping systems based on rainfall and soil characters

Rainfall(mm)	Soil type	Suggested cropping system
350-600	Alfisols, vertisols	Single rainy season cropping sorghum / maize/ soybean
350-600	Deep aridisols, entisols	Single cropping of sorghum / maize/ soybean in *kharif* / *rabi*
350-600	Deep vertisols	Single cropping, post rainy season *i.e.* sorghum
600-750	Alfisols, entisols	Intercropping, Sorghum + Pigeonpea, Cotton + Blackgram
750-900	Entisols, deep vertisols, deep alfisols, inceptisols	Double cropping with monitoring Maize – safflower, Soybean – chick pea, Groundnut - horse gram
> 900 As above	-	Assured double cropping Maize – chick pea, Soybean - safflower

Practical Excercises 331

Exercise No. 19: Visit to dryland research stations and watershed projects.

Visit of student can be arranged in the nearby watersheds. Few watersheds situated in Haryana are as under:

1. **Bunga watershed:** The project having an area of 463 ha is located at village Bunga in district Ambala.The average annual rainfall is 1115 mm. The soil is loamy sand and sandy loam with gravels on the surface.

2. **Bajar-Ganiyar watershed:** The project is having an area of 349 ha in Bajar and 746 ha in Ganiyar and is situated in district Mahendergarh. The average annual rainfall is 640 mm.

3. A part of Shivalik which forms foothills of Himalayan ranges also lies in the district of Panchkula, Ambala and Yamunanagar of Haryana state. This area, 1.92 lakh ha forms part of five watersheds namely Sirsa, Ghaggar, Dangri, Markanda and Yamuna. This is further divided into 16 sub watersheds and 162 micro watersheds. The area was taken up of development under IWDP in the year 1990 and lasted upto1999. There after it continues as Phase II. The main objective of the project is to restore the productive potential of the project area by developing osil-plant-water resources of the watershed to rectify the man made ecological imbalance by conserving the natural resources on arable and non arable lands.

A visit will be arranged for the students to these watersheds for on the spot study of various ongoing programmes in the watershed area.